Buntys and Pinkies
CHRONICLES OF A NEW CANADIAN

Canadian cataloguing in publication data

Chandwani, Ashok, 1950 -

    Buntys and Pinkies : chronicles of a new Canadian

    ISBN 1-895854-65-2

    1. Immigrants - Canada. 2. Emigration and immigration - Social aspects. 3. Emigration and immigration - Psychological aspects. I. Title

FC104.C44 1996   305.8'00971   C96-940798-X
F1035.A1C44 1996

The complete and ever-evolving
Robert Davies Publishing catalogue
is available on the Internet at :

http://www.rdppub.com

Buntys and Pinkies

CHRONICLES OF A NEW CANADIAN

A selection of columns
by Ashok Chandwani
of the *Montreal Gazette*

Edited by Lucinda Chodan

ROBERT DAVIES PUBLISHING
MONTREAL—TORONTO—PARIS

Copyright © 1996, Ashok Chandwani
ISBN 1-895854-65-2

ROBERT DAVIES PUBLISHING,
311-4999 Saint-Catherine Street, Westmount, Quebec, Canada H3Z 1T3
☎1-800-481-2440 / ☎ 1-514-481-2440   1-888-RDAVIES

Distributed in Canada by General Distribution Services
☎1-800-387-0141 / ☎1-800-387-0172   1-416-445-5967;

in the U.S.A., from General Distribution Services,
Suite 202, 85 River Rock Drive, Buffalo, NY 14287
☎ 1-800-805-1083

For all other countries, please order from publisher.

e-mail: rdppub@vir.com
Visit our Internet website: http://rdppub.com

The publisher takes this opportunity to thank the Canada Council
and the Ministère de la Culture du Québec (Sodec)
for their continuing support of publishing.

For Lynn

Table of contents

Introduction..............9
So dear and so far (Dec. 9, 1991)..............13
A gleam, please (Dec.16, 1991)..............16
Home sweet home (Dec. 23, 1991)..............19
Racism's tiny traces (Feb. 3, 1992)..............22
Expo imports (April 27, 1992)..............25
Anti-nuclear moves (June 15, 1992)..............28
Learning about lox (July 27, 1992)..............31
Dial J for jungle (Aug. 3, 1992)..............34
Labelled for life (Aug. 17, 1992)..............37
Scientists gone mad (Aug. 31, 1992)..............40
A ringside seat (Sept. 28, 1992)..............44
Good cop, bad cop (Jan. 14, 1993)..............47
Soldiering for love (Feb. 8, 1993)..............52
Salaam alikum (Feb. 22, 1993)..............55
Odious in any clime (March 15, 1993)..............58
Easy way in – or out? (June 21, 1993)..............63
Politics of survival (July 26, 1993)..............66
Dreams, just dreams (Sept. 20, 1993)..............70
Family? What family? (Nov. 15, 1993)..............74
A matter of respect (Nov. 22, 1993)..............78
Riches, here I come (Nov. 29, 1993)..............82
Looking for bonds (Dec. 13, 1993)..............86
'Tis the season (Dec. 20, 1993)..............90
Thanks, Miss Francis (Jan. 3, 1994)..............94
The rules of Rome (Jan. 24, 1994)..............98
A two-edged sword (Jan. 31, 1994)..............102
Running spooked (Feb. 28, 1994)..............106
Cause for despair (March 28, 1994)..............109
There's always hope (April 25, 1994)..............113
Africa's miracle (May 9, 1994)..............116

Fierce expectations (Aug. 1, 1994) 120
A turn with Mama (Aug. 29, 1994) 123
Buntys and Pinkies (Jan. 2, 1995) 126
A lousy year (Jan. 9, 1995) 130
Home-cooked joy (Jan. 16, 1995) 133
Airborne's legacy (Jan. 24, 1995) 136
A confusing battle (Feb. 6, 1995) 140
Terrorism's toll (April 24, 1995) 144
Perils of denial (May 8, 1995) 147
Roots of paranoia (May 29, 1995) 150
Trail of rejection (June 26, 1995) 153
Tears, just tears (July 3, 1995) 158
Looking to connect (July 31, 1995) 162
Moving's a party (Aug. 7, 1995) 165
Summer games (Aug. 21, 1995) 169
A facelift for Betty (Sept. 18, 1995) 172
Nightmare for all (Oct. 2, 1995) 176
Looking for labels (Oct. 23, 1995) 180
That sinking feeling (Oct. 30, 1995) 183
Room for all (Nov. 6, 1995) 186
Brutal truths (Dec. 4, 1995) 189
Pangs of bondage (Jan. 15, 1996) 193
A flag's just a flag (Feb. 19, 1996) 197
A stranger to all (March 11, 1996) 200
As the wheels turn (March 18, 1996) 203
What price tradition? (April 1, 1996) 206
Choo-choo magic (April 29, 1996) 209
Confession time (May 27, 1996) 212
On the margins (June 3, 1996) 215
Time for tea (July 29, 1996) 218
Bowing to the right (Aug. 5, 1996) 221

INTRODUCTION

What are Buntys and Pinkies?

On a snowy winter evening 23 years ago, the Turbo train brought me to Montreal from Toronto. A muddy bus with ice-crusted windows left me shivering on an Outremont street covered with knee-deep snow. The leather coat that had seemed so warm at the Army and Navy store in Vancouver barely four months ago now felt thinner than the loose cotton *kurtas* I used to wear in India. As the wind howled and blowing snow obscured street lamps and numbers, for the first time in my four months in Canada I found myself asking *Why*? Why was I here? It was a question that would surface many times in the following decades.

They say that time eases the pain of migration. They lie. Any immigrant who arrived as an adult can attest to this. It doesn't matter whether you left voluntarily or were forced to by politics or war or famine. In the new country you always remember the old. In the new kitchen, you simmer the old sauce. And for your newborn, you find names from the old home. Names like Bunty or Pinkie, if you happen to be from a middle-class Punjabi family. Or Joey or Luba or

Costas, for that matter.

Like these names, some of the baggage immigrants cling to is difficult to explain. Often there is no explanation – they're a tradition whose origin is lost in time. Other immigrant baggage is easier to understand. The way you dressed back home. The way you walked and talked. How you married and *whom* you married. Whether you dated. Whether you went to a church or mosque, temple or synagogue.

Immigrant baggage is rarely so strange that it's objectionable. Usually, it's just different. But that's not how it's always perceived in the new country. People there have their own baggage. Inevitably there is conflict and suspicion. All sides seek refuge in what's familiar, creating cultural and social ghettos. People use labels based on birth, gender, religion, skin color, sexual orientation and speech patterns. Hyphens are used to define nationality, which is often confused with race, ethnicity and language.

This leaves many of us reaching for a middle ground, a patch immune to distrust and prejudice. An Every Man's land that celebrates differences and rejoices in and nourishes diversity.

Mapping the road to such cosmopolitanism is one goal of the columns you'll find in this selection, culled from more than 200 that have been published in the Montreal *Gazette* since Dec. 9, 1991.

Other goals of these columns are to examine the deeper challenges of migration – those of dislocation, divided loyalties, identity, changing cultural values and acceptance in a new land.

Identity, nostalgia about childhood, the death of a parent – these are themes that strike responsive chords in all Canadians, not just the five million of us who are foreign-born.

I hope you agree.

Ashok Chandwani
Sept. 23, 1996

Acknowledgements

My deepest debt is to Lucinda Chodan, a colleague and friend, who edited this book. She read the first column before it became a weekly feature and offered invaluable editing advice. She has been generous with similar advice since, helping immeasurably in shaping the voice in this book.

For reading the first column, inflicted unexpectedly on him over an Italian lunch, and ensuring its birth as a weekly fixture, I would like to thank Norman Webster, former editor-in-chief of *The Gazette*. My gratitude also extends to other *Gazette* editors, particularly the unsung heroes on the night desk, for their patience and support.

I am also grateful to dozens of friends and family who have influenced the thoughts and stories in this book. These include Malika, Ajay, Sheila, Gokal, Gangu, Kalan, Michael, Jackie, Rabi, Kuki, Father Lawrie, Narendra, Mimi, Nandini, Zahid, Cynthia, Julius, Anuradha, Laurence, Maggie, Mina, Pandi, Shakuntala, Bill, Lindsay, Lisa, Ann, Brigitte, Jamie, Greitja, Garry, Wendy, Andrew, Hana, Satan, Sal, Donna, Bryan, Ray, Alan, Luci, Cooke, Maria, Penelope, Vian, Nat, Linda, Sanadee, the Cowboy, David, Alfie, Aylen, Sheree, André, Hubie, Michelle, Patti, Medea, Giancarlo, Jasmine and Shannon.

So dear and so far

She died on a hot afternoon in June in Bombay that was as muggy as the same morning in Montreal.

The static on the phone, 10,000 miles and 10 time zones away, yielded to a warm voice with a chilling message.

"I'm calling on behalf of your brother to tell you your mother died an hour ago. He's still at the hospital. He called the office to phone you. I am sorry."

> All immigrants leave family behind. When a parent dies back in the old country, the pain in the new country is compounded by the logistics of going home for the funeral.

My poor brother, I thought. Wasn't it yesterday that he had called about our father's death, same time, same city? No, it was three years ago.

Wasn't it yesterday I had phoned our mother? Cancer in remission, she had talked happily about cooking okra for my brother.

No, it was two days ago.

And now she was gone. An ache. An ambulance. A phone call to Montreal.

I'm glad you're home, she had said a year ago, twiddling

the controls on the tiny TV beside her hospital bed. An IV tube was taped to her forearm, drip-dripping chemo into her. I wish you weren't divorced. I wish your brother would get married. I wish you hadn't left home. . . .

"I wish I could help you," said the diplomat on the phone from the Indian High Commission in Ottawa. "But I can't fax you a visa. It has to be stamped in your passport. You have to send it here or bring it in person."

The joys of Canadian citizenship. A new passport. New rules and a new home.

The body would be ready for cremation soon, long hours before I could possibly get there. Friends would have helped my brother call relatives, find a priest, arrange for the funeral. Hindu funeral rites begin at home, last 12 days. Cremation takes place within hours; the eldest son lights the funeral pyre, the next eldest if No. One isn't around.

Three years ago I had arrived 36 hours too late to an urn full of my father's ashes and a red-eyed mother draped in mourning white.

Who would I find on this trip?

Your ticket will be ready by the time you return from Ottawa with the visa, my travel-agent friend said.

In slow motion I packed a bag. The plane would leave that night.

The phone rang again. Word must have spread. But it was my brother, home from the hospital.

Don't come home, he urged. It'll all be over by the time you arrive. I'm packed, ready to leave, I said. Think about it, he said. It'll cost a lot. Besides, I'll be at work and you'll be alone all day in the flat. You may as well mourn at home.

That's when the tears broke. Loud, wracking sobs. I called friends for advice.

It's good you're crying, said one, but you really should go home. It's only proper.

Stay home, advised another. What can you possibly do

arriving 36 hours later?

It's that call from hell, said a third. My mum's in England and I hate it when the phone rings in the middle of the night.

My husband's mother died in France and he couldn't go home for the funeral. He's never forgiven himself, said a fourth.

Awash in tears, I thought of other friends miles from their parents. Some from Australia. Others from Africa. The burden of all first-generation immigrants caught between continents. Between time zones. Between countries and cultures. Between homes.

I decided against going. I took three bereavement days. A friend drove down from Ottawa to help mourn. Others called, brought books, offered solace. A neighbor read and listened to poetry from the Gita and the Upanishads.

Bolstered, I returned to work. How comforting, I thought, to have such friends, such support. Home, I thought, is where your friends are.

A wait for the elevator opened the wounds again.

I'm sorry to hear about your mother, a colleague said. Did you go home for the funeral?

A gleam, please

The man spoke in mid-handshake, ignoring my polite "How are you?" at being introduced.

"Where are you from?" he asked.

In that crowded kitchen, during a noisy party, I found myself going still, brain switching to sensor mode.

It's a scan system many new Canadians develop, particularly those with "different" names or visages. It's primarily a defence mechanism – we're so eager to be accepted that we bristle at suggestions we're foreign.

> We all like to know where someone is from, be it Toronto or Trois Rivières, Montreal or Moose Jaw. But sometimes curiosity can take unexpected twists.

Most of us have torn up the old country's passport and proudly acquired the new one.

Most of us have been here long enough to feel some emotional involvement in the new country without necessarily losing all ties or connections with the old one.

Sadly, some of us "different" Canadians have been here several generations and still get asked where we're from.

Don't we appreciate genuine curiosity or interest in us?

Of course we do. Which is why we have this scan system. The questions that flash upon that inward eye are simple. The answers aren't.

Friendly gleam in the eye? Open your heart.

Hesitant look? Gingerly, now.

Suspicious look? Perverse response.

The last has a fine tradition drawn from answers to another question that used to bug new English-speaking immigrants.

Newcomers from Commonwealth countries, many of them accepted because they were university-educated professionals, would react with irritation when asked: Where did you learn to speak English?

To them the question reflected ignorance.

Many would try to make light of it.

"Oh, I learned it on the boat coming over."

"I picked it up on the plane."

But at least one doctor I know found herself thoroughly disconcerted when a Vancouverite responded with admiration: "My, you are *clever!*"

There were those, of course, who chose to be understanding and stated the obvious: that if you're a doctor or an engineer or other professional from a Commonwealth country, chances are excellent that you learned English from a young age. It might even be your mother tongue!

Today, with global telephone and cable-television links, such a question is rarely asked. Good riddance.

But there's still the problem of answering the one about origin.

"Why do you want to know?" I asked the man at the party, my sensors having registered hostility.

He was startled. He wasn't expecting a counterattack. But he wasn't going to back off.

"I'm just curious."

"OK, I'm from Outremont."

"I didn't mean that. I mean where are you really from?"

"Well, I came here from Port Perry, Ont."

"No, no, I mean where were you born?"

"Look," I said, smiling. "What does it matter where I was born? That's not the first question *I* asked you, was it? Why don't we chat a bit, have a drink, get to know each other? Then I can tell you where I was born, who I'm married to, divorced from, talk about my sex life. . . ."

"Man, you're sensitive," he said and turned away, leaving me with more difficult questions.

Was I being oversensitive? Why was it so important for this man to know where I was born at the very moment of meeting? Did he somehow feel threatened? If yes, why? Or was he indulging in a familiar urge – the need to label? To hyphenate? But without even a hi-how-are-you first?

I felt like a total stranger. And yet how many times have I cheerfully answered the same question from others, when the sensors positively glowed in the warmth exuding from the person asking?

"Don't let it get you down," laughed a friend with whom I shared these thoughts later.

"Look at me. I'm black, I was born here, my parents were born here and what's the question I'm asked most?

" 'What island are you from?'

"But I have the perfect answer for them.

"Montreal."

Home sweet home

H*alf to forget the wandering and the pain*
Half to remember days that have gone by
And dream and dream that I am home again!
– James Elroy Flecker

It's the ideal home away from home. The doors are always open, but there's no pressure to enter or stay.

Sometimes visitors arrive from Argentina or Australia. Occasionally they come from home, the old home, in the village of Ielsi in Campobasso in southern Italy.

Felicia and Pasquale Mirolla hold modest court here with endless supplies of food and homemade wine.

Turn up at 2 in the morning with sons Mike, Sal or Nick (or alone) and the welcome is the same as at 2 in the afternoon.

Many are the Sundays I've spent in the dining room, whose walls are covered with family photos.

Ma Mirolla, deceptively stern-looking, fussing in the kitchen. Papa Mirolla rushing downstairs to the cantina to siphon wine from demijohns into serving bottles that once held vodka or some meaner spirit. His own decanter bottle has a red-vinegar label on it – "so nobody else touches it."

In winter, lunch is indoors, sunshine streaming through frosted windows, a canary sometimes in song. In summer, it's under a canopy of grape vines, near Papa's pear tree.

There are always children, of all ages and many of them. Someone is always dropping by – brothers, sisters, aunts, uncles, cousins, girlfriends and boyfriends.

Sometimes visitors arrive from Argentina or Australia. Occasionally they come from home, the old home, in the village of Ielsi in Campobasso in southern Italy. Of late there is also Rudy, a talkative collie.

Attendance peaks at Christmas Eve, when most of the clan comes to Ma's for a traditional dinner.

I remember my first Christmas Eve with them. Twenty-four, married a scant year and a mere 15 months in Canada, the last three in Montreal.

My mother wants you to come home for dinner, my new friend Mike had said. You have no family here.

Simple as that. Everyone needs a family at Christmas.

The informality wasn't startling – I was born in a culture in which people routinely invite strangers home for a meal. But in a new milieu, it was comforting.

So on a snowy night, some 25 of us gathered in the Mirollas' basement. Festive tables with platters of olives and crusty bread. A mound of gifts under a decorated tree. Gallons of wine.

In keeping with a certain Roman Catholic (and family) tradition, it was a meatless dinner – one that Ma learned to cook at the age of 14 and that hasn't varied much in this century.

There was salted cod soaked three days, cooked two ways, baked and deep-fried in a yeast batter; cauliflower fried in the same batter, rayfish pickled in vinegar and saffron, rapini garnished with garlic and olive oil, spaghetti in an anchovy sauce and baked eel.

The children had their own long table, the younger ones

staring unabashedly at the gifts, the older ones feigning disdain for them. There were toasts and much ribbing. And at midnight, someone put on a Santa mask to hand out the gifts and put the children out of their misery.

Seventeen years have gone by since that first dinner. There have been marriages and divorces; some family feuds settled and new ones started; houses bought and sold; and some of the older children at that first dinner have become parents.

But they still keep gathering at Ma's.

Gearing up to cook a big Christmas dinner for the 56th time, Ma found time to reminisce. (Just this fall she had hugged goodbye a visiting sister from Argentina, a tearful farewell underscored by the unspoken thought that they would probably never see other again.)

She talked about life in Ielsi. Strict priests and fasts before Mass. The war and Papa's conscription. The frequent festivals on feast days. Their emigration. And bygone Christmases.

"In Ielsi," she said, "no one ever put up a Christmas tree." The tree, a northern tradition, entered their lives in Canada – the family album has a photo of their first tree-Christmas in 1952.

"And there was no Santa in Ielsi," Ma continued, "and we never exchanged gifts at Christmas.

"The only gifts we gave were little chocolates or something small for children on Jan. 6, the Epiphany.

"The night before, the children hung a stocking near the fire and in the morning they found a little something in them.

"But on Christmas Eve, no gifts. Nobody ever gave me anything on Christmas Eve."

A pause. And a twinkle.

"Except once. I got a ring on Christmas Eve. From Mr. Mirolla."

That was in 1938.

ASHOK CHANDWANI

Racism's tiny traces

'**H**ave you ever been the victim of racism?"

It was an innocent question, raised in the middle of a dinner discussion about race relations, a recurring subject these days.

I was among dear friends – guests at my table – but I found myself instantly uneasy. Expectant faces seemed to become one wet, candle-lit haze of espresso cups and wine glasses.

> Minorities are often accused of being oversensitive. But how would you react if people preferred to stand in a crowded bus rather than sit on the only empty seat beside you?

My friends were from varied ethnic backgrounds, but at this dinner, they were all white. Yes, I had been a victim of racism, but I wasn't sure they would understand, never having been victims themselves.

Racism often occurs in isolation, in one-on-one encounters. For every act of overt racism, there are 10 or more covert ones. Circumstance, nuance, sensitivity, intent – all come into play.

Take a female doctor who is dark-skinned. She's rushing at 2 a.m. from home to be beside a patient giving birth in a hospital. No white coat or stethoscope yet. In the hospital grounds, the cabbie starts veering toward the service entrance.

"The front, please," she says.

"Sure," says the cabbie, "I didn't realize you were a nurse." Double whammy. Surely if you're dark-skinned and female you must be in maintenance. All right, maybe a nurse. Certainly not a doctor.

Hard to believe? I know someone this has happened to. And she knows others.

I know an actress whose photo and CV were filed under Dwarfs and Exotics by a theatrical agent who didn't quite know what label to give a visible-minority performer.

I know of people who have been spat on, shot at or beaten; they've been ostracized, had homes vandalized, been denied jobs, housing and entry into public places because of their color.

OK, we minorities sometimes read too much into certain situations, but the overreaction and misunderstanding are usually the result of past abuse.

That supermarket cashier who fails to say thank you is probably not a racist, but is it just coincidence that the six preceding customers were all thanked and were all white?

The only empty seat in a bus is next to a black passenger, but *all* the standing white passengers don't feeling like sitting down? Another coincidence?

Do I sound angry? I'm not. Just weary.

This is Black History Month, and with all the legitimate pride and heritage on show, the sad truth is that after hundreds of years, the real issue for blacks remains racism.

We're supposed to recognize and acknowledge black history and culture in North America, but all we really acknowledge is that racial harmony is as elusive as ever. That white society is still reluctant to accept blacks as equals.

Yes, there is progress. But it's at the personal level, in your own social circle, your private ghetto. And at the legislative level – on paper, all those relatively new laws look swell.

But between the personal and the legislative lies a chasm that shows little sign of narrowing.

The dedicated organizers and participants (not all of them black) of the seminars, films, exhibitions and lectures this month in Montreal and other major North American cities will keep doing their share. And all power to them.

But, as Martin Luther King, the inspiration for Black History Month, would agree, the real answers lie in our hearts.

It was touching, that question from my dinner guests. Devoid of prejudice themselves, they couldn't imagine it in others.

Yet it wasn't really necessary to relive painful experiences for them.

They could see the answer written all over my face.

Expo imports

T*hose were the days, my friend*
We thought they'd never end
We'd sing and dance
Forever and a day
   *– **Popular song**
   by Mary Hopkin*

A quart of beer cost 65 cents, a glass 15, and orange juice was a nickel. You could gorge on a pepper steak for $1.95. And for the cab ride home the drop rate was 35 cents.

Expo 67 brought the world to Montreal. The story of three young men from Delhi is the story of thousands who experienced the city's open doors and open hearts.

Home for Parkash Oberoi, Yog Tandon and Ashok Kumar was a bed in a South Shore apartment building, where they were housed with the 22 other young men who were cooks, busboys and waiters at the Maharani restaurant at the Indian pavilion at Expo 67.

"There were five of us in each unit," recalled Tandon, sharing a few memories with his buddies on the 25th anniversary of the opening of Expo 67.

"We never got to bed before 4 a.m. and we were always scrambling in the morning. Often four of us would be in the washroom, one showering, one shaving, one brushing, one...."

"Reading the newspaper!" interrupted Kumar, chortling.

A Hungarian cabbie had been contracted to pick up the staff every morning. First the cooks and then the waiters, all for a 10 a.m. start.

"He was like a father to us," remembered Oberoi. "He knew we had been out drinking and dancing and he'd be very patient. Wake up, sonny, he'd say, or dear son or dear friend...."

"Drinking and dancing!" said Tandon, a bit dreamily, 50 now and father of two.

"Remember the girls? They came from all over. From Boston and Toronto and New York. We'd serve them lunch or dinner and then we'd make a date for midnight and La Ronde. And then...."

"Not me! Not me!" said Oberoi, also 50 and father of two, and still the pious one of the three friends. "What'll my wife think if she reads this?"

But then Oberoi met his wife at Expo 67.

"She was 18 then and was at the restaurant with her mother and brother. I thought he was her boyfriend, so I didn't care much at first. I didn't speak any French and she didn't speak any English, but my mother-in-law did."

Communication was never a problem at Expo. There were people from all over the world and Montrealers were eager to know them.

"They were so open," said Kumar. "Not like today, when many new people don't feel so welcome.

"Then, we all tried to learn each other's language and customs. I would show them how to do a folk dance and people would teach me new things."

Like Oberoi, Kumar also met his wife at Expo, is also 50, a father and integrated into francophone society. He's out of the restaurant business, but not his pals. Tandon owns Yogi,

BUNTYS AND PINKIES

an Indian restaurant downtown, and Oberoi is his suave and dapper maitre d'.

When Claridges, the Delhi hotel where they worked, got the contract for Expo 67, they were among the first to volunteer. For adventure, but also for money.

"Our monthly salary in Delhi then was 150 rupees, at Expo we got $100 – six times as much, plus tips!"

But the only time they could spend the money was in the wee hours. They worked seven days a week, from 10 a.m. to midnight.

"The restaurant was very popular," said Tandon. "There'd be a lineup at 10. Every night we'd turn people away.

"We served U Thant and Jackie Kennedy and Princess Margaret. Roland Michener was a regular and so was Mayor Drapeau."

When they got a short break each afternoon, for an hour or so, they'd be so tired from the night before they'd just sleep on the grass near the water.

"We never really saw much of Expo," said Kumar, "until later, when it became Man and His World."

None of them came with intentions of staying. Even when Oberoi and Kumar fell in love and decided to marry, they planned to return to India. Kumar actually did, but briefly.

Any regrets?

"None," said Kumar. "We made a decision and we're happy. Life is always up and down, but I have no regrets."

"I'm very content," said Tandon, who veered into anglophone society. "My sons are at school and university. And they're perfectly bilingual."

And how would they describe themselves now? As Canadians?

Thoughtful nods from Tandon and Kumar and a joyful outburst from Oberoi:

"I'm *French* Canadian now!"

Anti-nuclear moves

We got up at dawn that summer morning, my grandfather and I: he calm and unperturbed as ever, I agog with excitement. This was the day I would learn to swim.

We walked down to the river, one hand clutching my grandfather's, the other carrying a makeshift float made from an empty tin drum of shortening welded airtight.

As life expectancy and health-care costs rise, the traditional extended families many immigrants left behind are making a comeback in the New World.

I can still see myself, splashing nervously in the morning sun, the drum strapped to my back keeping me afloat, my grandfather's arms propping me from below, learning the first strokes.

The memory is as vivid as the one of my grandmother, in a different town in India, sitting beside my pillow, under a large mosquito net, ruffling my hair and crooning ancient folk songs designed to appease the pox goddess – and take a child's mind off the itchiness of chicken pox.

But in later years, other contrasting memories took root: the same grandparents, so adorable as visitors, turning into inflexible monsters as permanent residents.

We were all trapped in tradition. It was my father's duty, as the eldest son, to look after his parents. It was my mother's duty, as his wife, to do the actual looking after. I remember the tears on both sides, impotent ones of age from my grandmother, suppressed ones of rage from my mother. Over such silly things as the *amount of salt* in a curry. And the pampered, hypocritical men dismissing it as a *women's squabble*.

Yet, as with most of their generation, it never occurred to my parents to put my grandparents in a home. For starters, there weren't that many around.

It was equally unthinkable, years later, for my parents to live anywhere else but with my brother in Bombay, their eldest son having fled 10,000 miles to Canada, where such problems seemed to have been tidily resolved by a network of senior citizens' homes.

Not that there weren't any homes for seniors in Bombay by then. As in other developed or rapidly developing nations, increased urbanization had led to their proliferation.

But traditions in older, long-established societies die slowly. And they often tag along with their people, even if they move to newer and younger societies. Such is the heavy baggage of migration and transplantation.

Which is why, even today, there are Canadians of Indian, Chinese, Cambodian, Caribbean, Italian and Greek heritage (by no means a complete list) who wouldn't dream of putting their parents in homes. Or, conversely, tell their adult children to find their own apartments.

Which is also why many new homes today have self-contained sections for parents or in-laws and 30- and even 40-year-old sons and their families living at home with their parents, blurring the question of who's looking after whom.

But in the larger Canadian society of nuclear families splintering further into single-parent ones, the clock is turning a full and ironic circle. The population is aging and living longer, creating tremendous cost pressures on social programs. And governments are looking for solutions.

Quebec has come up with an interesting approach – a refundable tax credit for people who keep their "direct ascendants" at home.

It's not a large sum – $440 annually for each parent, grandparent or in-law over 70, or over 60 if suffering from a disability.

But it's not the amount that matters, it's the concept. I doubt that the Quebec government was thinking of its recently acquired and growing population of new Canadians in offering this credit. For the tradition-bound among these, the $440 will come as a minor windfall – perhaps a weekend in New York or a new parka and boots in winter.

No, as the budget document indicates, the government, while concerned about rising costs, is also hoping that the credit might restore traditional family support for the mainstream elderly: a socially satisfying role that has eroded in the postwar economic boom.

It seems the writing is on the wall everywhere, though the answer is hardly that found in Tokyo today, where there are agencies that will send elderly parents a trio of actors to act as a surrogate family of visitors for a few hours.

About $1,500 buys you three hours with a fake son, daughter-in-law and grandchild, paid for by the real relatives who are too entangled in the rat race to spare time for their elderly.

No, for my money, give me the the real thing, tin drum and all. And, if you must, the tax credit.

Learning about lox

During their visit (to Waldman's Fish Market) on Sept. 14, the inspectors said they saw several customers squeezing the fish on display.
– Excerpt from a Gazette report last month about a $500 fine levied by Montreal Municipal Court against the store for displaying products within customers' reach

For decades, a Montreal market offered customers fish from all over the world. Along with the fish came the smells and sounds of the Old Country, making it an early model for cultural integration.

When Waldman's Fish Market closed forever 10 days ago, victim of a labor dispute and a curious fine or two, a 68-year tradition died, too.

This was the store where newcomers found their favorite Old Country catches. Its fortunes were always pegged to the flow and needs of immigrants.

Max Waldman, who founded it in 1924, and his sons Benny and Morris, who ran it later, all knew that. Their deaths over the years – Morris died in 1987 – left a hole not only in their customers' hearts, but fears about the store's future that finally came true in the hands of successive corporate-style owners.

The best years at Waldman's were the 1960s and '70s, when liberalized immigration brought people from more than 100 countries to Montreal.

In those years, shopping at Waldman's was either a pilgrimage or a military-style operation.

Out-of-town visitors would gawk in disbelief and envy at not just the crowds of shoppers, but also the varieties of fish. Baby shark, mahi-mahi, grouper, snapper, bellyfish and monkfish by the heapful beside trays of more familiar haddock, cod, trout and salmon.

Then there were the tanks of lobsters, the mountains of shrimp and clams and mussels and oysters.

But it was one thing to go there on a Saturday afternoon with a visitor to savor the mayhem, quite another to actually buy fish.

In those days, the health inspectors weren't as picky. The fish, flown and trucked in from around the world, sat on the counters, which were always six or seven deep in customers.

"You want to take the fish home or just play with it?" a jovial fishmonger would ask, wiping his knife and hands on a blood-stained apron.

Embarrassed by the amused attention of other shoppers, a hesitant newcomer would stammer a request in a Portuguese or Greek or Bengali accent.

Others not so shy would poke around the mounds of fish, looking for familiar shapes and the telltale brightness of eyes of the freshest.

Yet others would be jostling, even wrestling, over the shellfish, elbow-deep, feet a little loose on the floor made slippery from spillage.

If you wanted to make a quick buy and exit, you'd have to go at a certain hour on a certain day, depending on the kind of fish you craved. Some only arrived once a week, others at a certain time of day pegged to the schedules of airlines and the trucks from Boston and New York.

If you wanted more than one kind, you had to take a friend or two along.

"You get the baby clams, George, while I look for the female lobsters."

Yes, this was where I learned to "fish" for a female lobster, prized for its roe, but indistinguishable from the male to the novice.

And in watching Bengalis buy bellyfish I learned it was the closest they could come to their beloved hilsa. In watching the Portuguese buy squid, I gained new respect for this otherwise ugly-looking species. Ditto with squirming eels in the hands of Italians and blue crab in the hands of Chinese.

This was also where lox entered my life, although with time and experience, there was better smoked salmon to be found elsewhere.

With time also, fish took off in Montreal, with new stores opening in other parts of the city, catering to greater demand and the immigrants who moved away from the traditional core on and near the Main.

Now Waldman's is no more, but for thousands of Montrealers it'll be a long time before the din and the babble, the vats of pickled herring and the acres of fish, will fade from memory.

Dial J for jungle

> Not long ago, the telephone was used only for business or emergencies in countries like India. Now, savvy insomniacs and homesick immigrants use it to track friends across time zones.

The way Zahid, a college chum in Bombay, described it, he was enjoying a post-lunch smoke on a verandah in the middle of a jungle in the western Indian state of Gujarat when the caretaker brought him an ancient telephone trailing a tattered cord.

" 'For you, sahib, from Canada,' he said. I almost fell off my lounge chair. Here I am in the middle of nowhere on a boring business trip and you call from Montreal.

"This technology business is getting ridiculous, don't you think?"

Indeed.

This technology business, notably in the field of telephones, has become the bane of migrants.

When I left India about 20 years ago there wasn't a hope of pulling a caper like calling this forest bungalow near the town of Rajkot. The telephone system just wasn't geared for it. Making a long-distance or "trunk" call required great planning and forethought. It wasn't uncommon for people to *write* and inform friends or relatives that they would be calling on a certain day so they would stay home all day.

You had to suffer an elaborate and time-consuming process – booking the call, getting a docket number and talking to hordes of operators – before making any voice contact with your party. Often the only people you talked with were the operators along the switching routes.

I remember making calls to my girlfriend in Vancouver from the central Indian city of Nagpur in 1973.

I lived in a student hostel run by the YMCA in those days. With the warden's permission I would set up a camp cot overnight in his office, beside the phone, to wait or sleep the eight or so hours it took to make the connection.

"HELLO! IT'S ME, I'M. . . . "

"SPEAK UP, INDIA!" the Canadian operator would shout. "WE CAN'T HEAR YOU!"

"SPEAK UP, NAGPUR!" the Indian overseas operator would scream.

"Speak UP, sir!" the local operator would urge helpfully.

"I AM SPEAKING UP!" I would bellow.

All that's changed now. You can direct-dial almost anywhere in the world from almost anywhere in India. The same is true, my friends tell me, for almost all the countries that once had poor phone systems. Simultaneously, it's also become far easier to call around the world from Canada.

This progress is doubtless a blessing for the thousands of new Canadians who jam the circuits to the Old Country on religious holidays or Mother's Day.

But it's really a curse for someone who has friends in a dozen cities and frequently feels the urge to converse with them.

And, as I always assure my friends, the timing of these urges is not related at all to bar-closing hours. It's more to do with time zones.

I mean, how many friends can you call in the same time zone at 3 or 4 in the morning?

But, hey, it's merely midnight or so in Victoria, B.C. ("Hope I didn't wake you up, Penny"), breakfast time in London ("Thought I'd catch you before you left for work, Andrew") lunchtime in Dubai ("What's cooking, Harpreet?") and mid-afternoon in that jungle in India where I called Zahid.

I tried him in Bombay first, and got redirected to Rajkot, only to find out that he had chosen the forest lodge over a hotel in town for peace and seclusion. That was the official version.

The real reason, he confessed, was that it was an irresistible way to goof off work for an afternoon.

"How often can you stay in a forest bungalow on business?

"But next time I think I'll ask them to unhook the phone!"

Labelled for life

Four years ago, in a dusty Indian square bustling with pilgrims, priests and peddlers, I came face to face with that most ancient and unsettling of practices – ethnic labelling.

My brother, our uncle and I had just driven into Nasik, a pleasant town on the banks of the Narmada River, about 200 kilometres inland from Bombay.

> On the banks of a holy river, I found a complete family tree built on branches of deaths. It was a tidy record of my past based on language, ethnicity and geographic origin.

Our mission was to immerse my father's ashes in the Narmada, one of several rivers considered sacred by Hindus. Religious ritual dictated that the final rite of immersion take place on the 11th day after death, in one of seven sacred cities scattered across India.

Our taxi, dodging people and cows in Nasik – the town's sacredness gave them divine right of way – had barely come to a halt when a dozen young Brahmans, dressed in white, heads shaven but for a mini-braid, surrounded us.

"What's the name?" they shouted in Hindi.

"Chandwani?" someone responded, switching to Sindhi. "You're Sindhis! You're mine!"

"We deal with Sindhis, Gujaratis and Rajasthanis," our captor continued as his competitors filtered away into the crowds.

"The others deal with the other states of India."

He produced a red scroll-shaped ledger from a shoulder bag.

"Follow me," he said, as he flipped through long, time-creased pages covered with neat handwriting in Hindi, found a lengthy entry and chanted:

"Chandwanis from the town of Ratodero, province of Sind, now in Pakistan. You must be Ashok and Ajay, sons of Gokaldas, and you must be Ganguram, brother of Gokaldas, son of Harumal, son of Tirathdas, son of...."

Imagine my astonishment. It was all here – a complete family tree, built on branches of deaths, a tidy record of my past based on language, ethnicity and geographic origin.

It was an astonishment that deepened when I questioned the young priest about his archival system.

All the priests have family or other connections in the other sacred cities and after each rite, they exchange information. Thus, ledgers in all the cities are constantly updated and there is talk now of putting them on computer.

But my astonishment was also tinged with irritation.

Here I was thousands of miles from Montreal, in a town I had never been to before, and someone seemed to know all about me.

Not only that. Without my permission, I had become the property of a family of priests whom I had never seen until that day. And had I been born, say in Madras in a

Tamil-speaking family, I would be the property of some other group of priests.

It didn't seem to matter to anyone that I was an agnostic. Or that I didn't live there. My name and my mother tongue were all that mattered.

I was marked for life, I thought, as I sat through the pooja (prayers) on the covered front porch of the priest's house in a crowded lane, steps from where the river bank is lined with stone platforms, pathways and temples.

You can be a refugee or an immigrant, never leave home or make a new one somewhere else. But you can't ignore your past and shed the baggage of centuries.

And, in a new country with a new culture and new rules, even if you succeed in dropping a few layers and adding new ones, there are people and forces waiting to remind you that your efforts aren't enough or that they are irrelevant.

Ethnicity, then, becomes your identity, with geography and numbers determining your clout in society.

For most in the mainstream, you always were and will remain a Sindhi, a Bengali, a Tamil, an Indian; a Campobassan, an Italian; Cantonese or Hong Kong Chinese; a Somali, a Greek, a Haitian, a Korean.

Anything but a Canadian.

An ethnic.

Scientists gone mad

Armed with a post-doctoral degree from France and fluency in five languages, Chander Grover arrived in Canada in 1978 to continue a career in fibre optics.

After a "struggle for a year," looking for suitable work, the Indian-born scientist found himself making rapid progress – in research and status at the National Research Council in Ottawa, which he joined in 1981.

> Men and women with advanced degrees appear to have spent hours scheming and managing the harassment of a leading scientist with a proven research and managerial track record.

Budgets, labs, research teams and promotions all fell into place smoothly as Grover moved from one project to another.

Then, in 1986, he turned down an offer to be scientific director of the National Optics Institute – an organization he had helped start and had been delegated to by the NRC.

His reasons were personal and simple. The new job would entail a move to Quebec City, disrupting his children's schooling as well as that of his wife Sumi, who was working toward a master's in information sciences, after having interrupted her doctoral studies in chemistry some years earlier to raise their children.

So Grover decided to remain in Ottawa and resume full-time duties at the NRC – for the two years previous, he had been spending 40 per cent of his time with NOI.

That's when everything fell apart.

That's when, according to a ruling six years later by a Canadian Human Rights Tribunal, Grover ran into "far from covert" racism from some NRC managers – racism that was "flagrant and calculated to humiliate and demean."

There had been some managerial changes and restructuring while Grover had been busy with the NOI. Back full time with the NRC, he mysteriously found his budgets being withdrawn, his team disbanded, his lab and office space reduced.

"Everything I proposed or did seemed to be undermined or rejected, without a reasonable explanation," Grover recalled on the phone last week. "There was no work for me to do."

The internationally recognized expert on fibre optics concluded that the only explanation for the harassment was his color. And he decided to complain about it. The complaint took almost six years to resolve and the Aug. 21 ruling, which could be appealed by the NRC, was also a severe condemnation of NRC managers.

According to the tribunal report, "the general treatment of Dr. Grover and the effect on his personal health and family life has been both demeaning and devastating."

The tribunal ordered the NRC to immediately cease and desist from its discriminatory conduct against Grover, promote him to the position of section or group head as soon as possible, pay him lost wages that could total thousands of dollars and thoroughly review its human-rights program.

The tribunal also ordered that the president of the NRC

write a formal apology to Grover within 15 days of the order.

And the tribunal strongly recommended that evidence presented at its hearings be turned over to the attorney-general for prosecution of some employees under the Human Rights Act. The evidence indicated that an NRC employee tried to conceal information and a Justice Department official tried to intimidate one prospective witness. The tribunal also criticized the "impropriety of conduct and obvious conflict of interest" of the NRC's own human-rights adviser in handling Grover's complaint of discrimination.

Excerpts from the tribunal's 94-page report and the phone conversation with Grover last week reveal a disgusting pattern of harassment at the NRC. It puts a repulsive face to a national, taxpayer-supported institution dedicated to scientific research.

Men and women with advanced degrees and training appear to have spent hours scheming and managing the harassment of a leading scientist with a proven research and managerial track record.

Typically, as the tribunal found, attempts were made to portray the victim as the troublemaker. The tribunal rejected this defence and another contention that Grover was treated no better or no worse than any other staffer.

Indeed, the tribunal went out of its way to praise the candor and credibility of Grover and colleagues who testified on his behalf, in an atmosphere of intimidation. And it described as evasive and uncertain the evidence of a key respondent, his supervisor.

"His (the supervisor's) manner of giving evidence was evasive, his conclusions totally implausible and his explanations, in particular for the way he conducted himself as Dr. Grover's superior, were inexcusable, insensitive and demeaning to a fellow scientist," wrote the tribunal.

All these events left Grover shattered. He found himself

depressed and irritable. He stopped going with his son and daughter for hockey and softball. And he wondered what effect his experience would have on the children, both above-average performers in school. (His son Samir is on track to complete Grade 12 at the age of 14, at least four years before most other students do.)

"But they were understanding and supportive," he said. "We sat and talked about it.

"And my wife was there at all 28 sittings of the tribunal, by my side all the time."

A firm "belief in myself" also helped him cope.

And does he feel bitter about it all? Five years of doing nothing and the time it will take to re-establish, even if there are no appeals?

"As I joked to some friends," he replied, "I'll probably retire five years later than I would have.

"I don't believe Canada is a racist society.

"I was just unlucky. I ran into a few bad apples."

A ringside seat

So which side are you on, Mr. or Ms. Ethnic Canadian?

The predominantly anglophone Yes or the predominantly francophone No?

For, putting aside the divisions in both groups, isn't this (however historically correct) once again a largely English-French exercise?

No one comes right out and says so, but the Yes side assumes allophones will vote with it. Even Jacques Parizeau must fear so, given his repeated statements that no one should make such an assumption, least of all the Yes side.

This places allophones, who might agree with Parizeau, and others not quite as docile and compliant as the Yes side sees them, in a tough spot.

Here we are, eager to at least listen to – if not engage in – intelligent debate about the Charlottetown constitutional deal and what do we hear most? Fear-mongering.

> Eager to participate in the Charlottetown debate, allophones, as Quebecers who are not from traditional English- or French-speaking backgrounds are called, find themselves on the sidelines.

First we have the St. Jean Baptiste Society launching an ad blitz that implicates "newcomers" for perceived threats to the future of francophones.

And then, the other day, we have Joe Clark on television looking solemnly into the camera and saying "Beirut – it could happen here."

Both the SJBS and Clark are wading into dangerous waters in raising such spectres. Both messages are guaranteed to alienate allophones, with Clark's being – arguably – the more disturbing.

Respected editorialists on both sides have already denounced the SJBS ads and fortunately the society remains a minority group, its capacity for greater harm limited.

But when the federal minister for constitutional affairs links a No vote to a collapse of law and order paralleling Beirut, you have to go beyond the predictable denunciations and wonder about the play here.

A lot of new Canadians have either fled such disorder and violence or are familiar with it because they or their parents lived through it.

You don't have to spend all that much time with someone with roots in Lebanon or the Indian subcontinent before the talk turns to the horrors of internecine conflict.

But it is these very allophones who will be the first to admit that the social or political conditions preceding the militia wars in Beirut or the partition riots on the subcontinent just do not exist here.

The religion-driven rifts and colonial power plays are absent and so is the despicable tendency to take up arms.

As an allophone who enjoys the streets and bars and restaurants of Montreal alone or with friends from the full spectrum of this city's residents, I just don't see even one-hundredth of the tensions and hostility required to create a Beirut here.

After all, Beirut didn't happen overnight and wasn't pegged to a single event. It takes time and blind hatred to gather those Uzis and Khalashnikovs, not to mention bazookas and rocket-launchers.

No, Mr. Clark, such conditions just do not exist and never will, notwithstanding the need to find a solution to our constitutional crisis.

We all are and must remain civilized, as must the tone and content of the unity debate.

Yes, there will be hot words. And tempers will doubtless fray. Some misguided lunatics might even exchange a blow or two.

But no one should decide which side he or she is on because he or she is frightened. Because, as every allophone is often told, that's just not the Canadian way, is it?

Good cop, bad cop

> Government officials can all strike fear in many newcomers. But being stopped by cops in the wee hours is an experience everyone dreads.

It was about 1:30 a.m. on a chilly night some weeks ago that a police car began following mine on a deserted street.

About a dozen stop signs and traffic lights later, the car pulled up alongside in the outside lane as I waited for a light to turn green.

By now I had become nervous. Why had they followed me? I wasn't speeding, going through stop signs or lights or driving strangely.

True, I had stopped at a bar on my way home from work – journalists work the same weird hours and shifts as doctors and cops. But a couple of drinks over a couple of hours was all I was guilty of.

Maybe it was my eyes-straight-ahead-I-know-you're-there-but-I'm-innocent look; maybe it was palpable nervousness at unexpected and unnecessary (in my view) police scrutiny; maybe it was the way I was drumming along to the radio, as two pairs of eyes raked my left cheek and neck.

But, sure enough, instead of turning left, the squad car

waited as I drove through the green, followed me and put on its flashers.

I had absolutely no reason on Earth to be afraid or apprehensive – I had nothing to hide, no parking tickets unpaid or any other record. Just Joe Citizen going home.

But in hindsight, I think I must have been suffering from what I will loosely label "cultural terror." It's the feeling that suffuses someone from a minority when confronted with mainstream officialdom, particularly the police.

It's akin to the feeling a traveller of any background might get not just at customs and immigration checks in foreign countries, but also at railway ticket or hotel counters.

It's the feeling, in our North American context, that many non-white people and organizations regularly refer to in discussions of racism. A feeling that's compounded in minority minds each time there's a fatal shooting by police and the person killed is someone of color.

In trepidation, I rolled down my window as I watched the two male cops approach in my driving mirror, hands and bodies positioned as you see them in the movies, alert for a hostile reaction.

As they came abreast, one closer to the car than the other, I noted the one "covering" his buddy was not white. I had known that the Montreal Urban Community has been trying to increase ethnic representation on its police force, but this was the first time I had ever actually seen a cop of Asian heritage.

I can hardly describe to you the sense of relief that began creeping upwards from my tense toes. Whether from conditioning or skewed expectations, I felt that some sort of ethnic balance had worked its way into the equation.

The conversation that followed only added to this relief

– some of which turned out to be comic.

The francophone cop, whom I'll call Cop A, did most of the talking, after politely establishing that I would be more comfortable speaking English.

He accused me of going through a stop sign a dozen or more blocks ago. I denied the accusation, politely pointing out that he had been following me for a while and surely must have noticed that I was obeying all the signs. Curiously, he did agree and then asked me if I had been drinking. I said, yes, I had had two drinks. Then he wanted to know if the car I was driving was mine, and then, where did I work?

The Gazette, I answered.

This is where the episode took on a surreal tone, the exact implications and conclusions of which are still unclear in my mind.

Ask him his name, Cop B suddenly muttered in French from his 2 o' clock position to his buddy. Cop A dutifully did.

"Page Two column on Monday!" exclaimed Cop B, much to my surprise, after I answered. I had no idea Montreal cops read *The Gazette,* let alone my weekly column.

"What?" exclaimed Cop A, as surprised as me.

Cop B: He writes in *The Gazette.*

Cop A (unsure and then facetious): The Gazette? Is that a good paper?

Cop B (a bit huffily): Well, I read it!

The whole atmosphere changed after that. Whether it was because I was a journalist, or whether it was the unexpected interruption from his buddy, Cop A gave up on the stop sign and returned to the subject of drinking, which only brought more unexpected remarks from Cop B.

Cop A: How many drinks did you say you had?

Me: Two.
Cop A: And what did you drink?
Me: Metaxa.
Cop A (sticking his head through car window, ostensibly to hear better, but obviously to smell my breath): What's that?
Cop B (from behind): It's a Greek brandy!
Cop A (sheepish, distracted): Oh yeah?

I almost thought he would ask whether Metaxa was a good brandy, but, his manner indicating puzzlement at his buddy's knowledge of ethnic columns and brandies, he resumed what had by now degenerated into a farcical interrogation.

Refusing a friendly offer from me to submit to a Breathalyzer test – it was getting close to 2 a.m., I wanted to get home and thought this would be a face-saving device to end the episode – he discovered from my papers that the validity of my insurance slip had expired.

Refraining from pointing out that he hadn't stopped me for an accident or an insurance check, I told him I thought I had put the new one with the papers, but if it wasn't there, then doubtless it was sitting on my desk at home.

Seizing his own face-saving device, Cop A announced that, regretfully, he would have to give me a warning ticket – I would have to produce the new slip within 48 hours at any police station.

The cops returned to their car to do the standard computer check on my papers and write the ticket, leaving me to sort out my confused thoughts.

Why, when all was said and done, had they stopped me? Clearly not for a stop-sign violation or they would have done so at the point of alleged occurrence instead of a dozen signs and lights later. Were they – frightening thought – just bored and wanted a bit of "action"? Was it because I drive a somewhat beat-up 17-year-old car? Or

– that feeling again! – was it the color of my skin? And finally, the most unsettling, how would I have been treated if Cop B hadn't recognized my name?

About the only comforting aspect of the encounter was the ethnic dynamic that affected it – from the nature of my column to the heritage of Cop B and his interaction not just with another ethnic, but also with his mainstream francophone colleague.

I hope, I started saying to myself, the encounter raises the same questions in the cops' minds when Cop B hailed me as he walked up with the ticket.

"Chandwani!"

Me: Yes , sir!

Cop B: What nationality is that?

Me (quelling urge to say "Canadian" in favor of quick departure): Well, I was born in India (and, unable to resist) – and you?

Cop B: I am Chinese.

Me: Good!

We exchanged smiles as he handed me the ticket. But he still had one more question, the easiest one of the evening:

Cop B: Tell me, if you are Indian, why do you drink *Greek* brandy?

Ah, Canada, I sighed as I drove home chuckling.

Soldiering for love

> Every culture develops its own dos and don'ts of romance and courtship. The fun (or should I say trouble?) begins when one culture encounters another.

A friend who once worked at the McGill University library tells the story of a pair of ardent brown eyes that she would suddenly find staring at her through cracks in the stacks.

"All I could see was these eyes and a bit of beard," she recalls.

The message in the eyes being unmistakable and unwelcome, she undertook the evasive measures women learn to take in such situations. But after weeks of nimble footwork, the inevitable happened.

Turning a corner, she found herself face to face with her tormentor – a South Asian student who stretched out his arms beseechingly and implored, "One night of love, just one night of love!"

"Go away, you beastly man!" my friend managed to shout, as she did an about-face and fled.

Some 25 years later, my friend looks back on the incident with amusement and some insight on the cultural nuances in romance.

"The poor man was probably alone and very lonely. And he certainly didn't get physically aggressive or show up in the stacks again.

"But I couldn't very well melt into his arms for a night of 'love,' could I?"

Most certainly not.

But as another Valentine's Day approaches, I can't help but muse about all those lonely hearts out there, some of whom doubtless belong to people uncomfortable, if not unfamiliar, with long-stemmed red roses and flutes of champagne.

Love is a universal phenomenon – no cultural boundaries here. But that's hardly true of its pursuit and practice.

Every culture develops its own dos and don'ts of romance and courtship. The fun (or should I say trouble?) begins when one culture encounters another.

Language, ethnicity, religion, dress, food, social customs and – regrettably – color all combine to create a stew of perceptions and misconceptions. Throw in something as intensely personal and subjective as romance and the brew can get complex, if not volatile.

Not all newcomers are looking for love and not everyone looking for love is first-generation. The cultural baggage that immigrants bring is durable – it stays around for several generations for most, forever for some. If you weed out the latter and ignore those among the former who have succumbed to the cozy mantle of similarity, you're still left with enough of the lonely and lovelorn who straddle cultures and languages.

It's not easy for these die-hard romantics, looking for love in this time of multiculturalism.

Naturally inclined or conditioned to both cherish and overlook differences, they find themselves tagged not just by the mainstream, but by other minority groups, too, for their particular distinctness.

Try as they might, these multicultural misfits who disdain being labelled to suit someone's stereotype just can't shake off that baggage from the past. Like Saran Wrap, it clings so viciously, you have to tear it off. Even so, a few strips might remain to torment you.

Undaunted, these forlorn souls soldier on.

They learn to pick the right color of roses and label of bubbly, talk knowledgeably about French Impressionists or Fassbinder, ski downhill or go bungee jumping, sip kirs or do upside-down kamikaze shots, curl, bowl or shoot pool.

Along the way, some of them strike the right note and romance blossoms. It's not uncommon these days to see intercultural couples, most of them young. And it's becoming common to come across names and heritages that are hyphenated not once but twice and thrice.

But somewhere in the equation you also find the recently arrived – most of them men, because of the economics of immigration. Legal, illegal or refugee claimants, the first ones over are generally male, the ones who work and save and send for the rest.

All have families back home – certainly parents or brothers and sisters, if not wives and children.

All are lonely, and Valentine's Day only makes them more so. It makes them mistake a friendly smile for romance, a friendly kiss or hug for a promise of giddy indulgence.

It makes them mumble about one night of love.

Salaam alikum

It's as predictable as snow in February – every so often someone assumes I'm a Muslim.

Now, the assumption in itself is not why I'm telling you this. As an agnostic, I don't really care what religion others think I am.

No, it's the stereotypes that create the assumption that disturb me. Profoundly so,

> The stereotyping of Muslims can sometimes take a bizarre turn for someone who has a brown skin and a 'foreign' name. Such ignorance only adds to the misconceptions about Islam.

because I end up wondering about how such stereotyping affects our society's attitudes and behavior not just toward Muslims, but other minorities.

To many Judeo-Christian Canadians I meet, I have a "foreign" name and vaguely Third World (as in brown) face. The face does not resemble what they consider to be a Latino face – so obviously I couldn't be sharing their religion. In which case I must certainly be a Muslim.

It generally comes as a relief to such people to find out that I am not a Muslim – a reaction that is also based on the stereotypical assumption that to be a Muslim is to be a terrorist or an Arab, or both. And also someone whose religion demands oppression of women. And they wouldn't want to associate with *that* kind of person.

So, in mid-conversation on unrelated topics, at parties or halfway through a romantic dinner, I find myself ranting a bit about misconceptions about Muslims.

This doesn't result in my listeners running off to the nearest mosque to embrace Islam, but it does cause some discomfort and, invariably, surprise.

It becomes apparent during such discussions that the main misconceptions afloat around us are that all Arabs are Muslims and that all Muslims hail from the Middle East or Iran.

Few pause to consider that the country with the largest number of Muslims (190 million) in the world is Indonesia, whose people bear no resemblance to the Semitic faces of the Middle East and whose women wear makeup and work as flight attendants.

Not many think of the hundreds of millions of Muslims in the Indian subcontinent and their distinctive sociocultural attitudes and practices. Or that, as in Turkey and Egypt, Muslim women in the subcontinent are broadcasters, engineers, professors, lawyers or architects.

Again, how many of us think of "Palestinian" as a synonym for "Muslim," when in fact, a third of Palestinians are Christians?

The point here is that there are a billion Muslims in the world, about 6 million of them in the U.S. and 350,000 in Canada, of widely different ethnic, linguistic, racial, social, political and artistic heritages.

The one thing these millions do have in common is their faith in Islam, a powerful religious and social force that has contributed enormously to human civilization since its birth in the seventh century.

Art, science, music, architecture, engineering, education

– there is hardly any sphere of human endeavor that has not been significantly affected, if not invented, by Muslims.

So what about those fundamentalists, you say, the ones who hand out death sentences on writers and mastermind and participate in terrorism in the name of Allah?

I say they deserve to be condemned as much – and as strongly – as their counterparts among Hindus, Jews, Christians, Sikhs and Buddhists. No human being, religious or nonbeliever, ought to approve of violence inspired by religion.

But, as Canada's Muslims begin a month of fasting and reflection this week, other Canadians should do a bit of reflecting of their own and realize that no human being should suffer for the misdeeds of others.

For Muslims, the holy month of Ramadan is a period of atonement, similar in that focus to Yom Kippur for Jews and Lent for Christians.

During this month, Muslims will abstain from food, drink and sex from dawn to dusk. And they'll spend more time in mosques for prayers and recitations from the Koran and for reflection.

The Muslims who will do this will be your neighbors and mine, fellow Canadians from a variety of backgrounds. Is it too much to ask that they're not tarnished by the actions or sins of a few or become victims of stereotypes they have no role in, knowledge of or relation to?

Isn't that what a secular, pluralist society is all about? The freedom to practise your religion without it becoming the basis for others' prejudice?

Odious in any clime

When my father and mother got married 51 years ago, in the thick of the Quit India movement aimed at colonial Britain, they did so in the traditional Indian fashion of a union arranged by their parents.

Astrologers picked the auspicious day and my father, dressed in handspun silk garments and sporting a turban of marigolds, led the bridegroom's procession to my mother's house on a white horse adorned with red and gold plumes.

> Newcomers always bring cultural baggage. But there are two things they should exorcise here and in the Old Country – the historical abuse and devaluation of women.

Under an outdoor canopy draped with flowers, as a priest chanted ancient Vedic verses and clouds of fragrant incense rose from a ceremonial fire, my father set eyes on my mother for the first time, though not on her face, which was covered by the end-piece of her red silk wedding saree embroidered in gold thread.

They only got to see each other's faces, my parents would recall shyly years later, after the ceremony in a nuptial bed also decked in flowers.

And, describing a marriage that would end when death did them part, they would proudly point out their rejection of a crucial component of their traditional marriage – the socially and mentally crippling custom of dowry.

Somewhat to the surprise and distress of their parents, my father had made his consent to his marriage conditional on eliminating dowry from the arrangements.

The decision, he would explain, had been inspired by the other nonviolent war Mahatma Gandhi had unleashed in India.

Freedom from external oppression by Britain, Gandhi had argued, should accompany freedom from the internal, non-British social evils prevalent in India, among them the shameful and shameless practice of dowry.

Thousands of young men and some women (their numbers were small for lack of emancipation, not desire) had responded to Gandhi's message, but, as we all keeping reading or hearing, their modest efforts aside, the progress made in eliminating this oppressive practice has been minimal.

The custom of dowry, in which a bridegroom's family demands and receives cash, jewelry and gifts from the bride's family, flourishes unabated and unchecked by legal restrictions, sometimes leading to murder.

The same Indian newspapers that report on and decry dowry-related bride burnings in India – incidents in which, sadly, mothers-in-law are usually involved – also print thousands of lucrative matrimonial ads that perpetuate the dubious aspects of the system of arranged marriages.

Inevitably, these abuses also migrate to other parts of the world as part of the cultural baggage of some Indians in much the same way other nasty habits of female genital mutilation tag along with some immigrants from certain African countries.

Last week, the Vancouver Sun and Province newspapers

reported on a disturbing case of dowry, arranged marriage and killing involving an Indo-Canadian family.

According to testimony during an inquest, the family of Jagtar Paul Dhillon in Aldergrove, 40 kilometres east of Vancouver, tortured and debased two women from India after bringing them, on separate occasions, to Canada for arranged marriages.

The testimony came from 23-year-old Interjit Toor, the divorced second wife of Dhillon, at the inquest into the death of his divorced first wife, Swaranjit Thandi.

The 25-year-old Thandi's beaten and strangled nude body was found on Feb. 3, 1991, at the bottom of an embankment near Merritt, about 200 kilometres northeast of Vancouver.

No one had reported her missing and no charges have been laid in the case.

Apart from the allegations of torture, abuse and slavery, other ingredients of this unusual case include testimony from the dead woman's father, who lives in India, about his futile efforts to extricate his daughter from the Dhillon family in the years preceding her death.

According to his testimony at the inquest, she was terrified to leave, fearing that he and her brother would come to harm from the Dhillon family.

The case is far from solved – a warrant has been issued for the arrest of Dhillon, who failed to obey a summons to testify at the inquest – and may indeed never be.

But to me, it highlights the sound reasoning behind the guidelines issued last week in Ottawa on the status of women who claim refugee status because they fear gender-based persecution in their home countries.

The issue is simple, clear-cut and universal. Certain countries, cultures and religious groups persist in continuing the historical devaluation of women. And the system of dowry, odious as it is, is just one of many gender-related persecutions women face all over the world, including in Canada.

Rape and physical abuse are the most talked-about ones; but there is also the issue of femicide, a chilling Asian practice that has lately got a boost from reproductive science.

Repugnant families in China and the Indian subcontinent, long accustomed to starving or killing outright their baby daughters, are using advances in prenatal sex-determination and selection techniques to eliminate women before they're even born – through abortions.

Consider these alarming statistics offered by Professor Vibhuti Patel from the Women's University in Bombay and Professor Gautum Appa at the London School of Economics, in a story in the *Guardian* in London last week.

In India between 1978 and 1983, 78,000 recorded cases of female abortion took place after sex-determination tests, and the sex ratio in one area went down in 1981 to 836 females to 1,000 males.

Reported annual dowry deaths in India have gone up from 358 in 1978 to more than 1,500 in recent years. A successful ad seen at some gender clinics carries the slogan "Spend 500 rupees now to save 50,000 later."

In South Korea, male births exceed female births by 14 per cent.

In one province of China, men between 35 and 40 outnumber women by 10 to one.

The sad truth is that the women who do survive into adulthood frequently suffer such abuse and deprivation that some of them start feeling it would have been better for them to have died as a fetus. (What a horrible thought!)

It is with good reason, therefore, that Ottawa officials and women's and other rights groups describe the refugee guidelines as a matter of adopting a trendsetting stand on universal values rather than an attempt to impose Western standards on other countries.

Rejecting the dowry system in response to social crusaders and guidelines on refugee claimants fearing gender persecution are part of an essential, albeit slow, process of world-wide social change that can only benefit all of us, no matter which country we live in.

Easy way in – or out?

For some new Canadians, the phone seems to ring for only two reasons – relatives calling to chat or to announce a death in the Old Country.

> Family ties can impose a special burden on newcomers – and lead to some uncomfortable situations. There's never an easy way out when someone seeks help to get in.

Having been through a few death calls in the past five years, it was with relief that I greeted a caller a few months ago from Chicago, a cousin whom I hadn't heard from in years – her voice sounded too cheerful to be the harbinger of bad news.

But what she had to say, after pleasantries, made me wish I hadn't answered the phone.

Unlike letters, telephones put you on the spot instantly. No time to digest and weigh – you have to respond instantly, often irrationally.

"Daddy asked me to call you about my brother. He's wondering if you can do us a favor," said my cousin brightly, putting my defences on red alert.

Her father, who is also my favorite uncle, had raised this

matter previously during a trip to India. His son, he had said, hadn't done well enough in school to get into college, making him unemployable. His daughter, who had married and moved to Chicago, wasn't completely legal yet and was unable to help her brother immigrate to the United States "where there are plenty of jobs, degree or not." Would I sponsor my cousin to immigrate to Canada instead?

I would if I could, I had replied, choosing my words carefully, but the laws were strict and favored immediate and dependent relatives. Besides, I said, Canada was not the United States, there was a severe recession, jobs were scarce or nonexistent, and my cousin would find it very difficult to cope, particularly in Montreal.

My aunt and he were desperate, my uncle, a retired petroleum engineer, had said, to "get their son settled." Perhaps I could make it seem that I was sponsoring my brother?

But I had only one brother, I had pointed out, and he had no desire to leave Bombay. I would be lying if I claimed my cousin was my brother. And what if my brother wished, one day, to join me in Canada, now that our parents were dead?

Oh, my uncle had replied, he could "arrange" for papers to show that I had a second brother. And what really was the difference between a brother and first cousin? We were all brothers! Sorry, I had finally said rather testily, I wasn't willing to sponsor my cousin and I wasn't about to get involved in illegal activities.

Back home in Montreal, I had had ample time to brood over my uncle's desperation to help his son – a desperation so acute that he seemed willing to consider breaking the laws of two countries. I felt guilty for having rebuffed what he obviously thought was a normal parental desire to do well by his children.

But even as I felt firmly that I couldn't participate in illegal schemes, I was painfully aware that I was doing it from a position of economic strength. I would have been happy to see the matter end there, but here was this unexpected call from Chicago.

It turned out that my uncle had concocted a new scheme that involved helping his son obtain a tourist visa to Canada. Apparently once here, he would make his own way to the United States – illegally, of course.

No way, I told my disappointed cousin, no way. I had already made it clear, I said, that I didn't want to be part of any such schemes, least of all illegal ones.

She didn't push it, but her call unleashed more disturbing, if familiar, questions.

How desperate could one be to come to the United States? Or Canada, for that matter? I could understand the motivation of people fleeing hunger, war or political oppression. But my cousin was hardly in that category. Was it merely for the good life, then? And was there really a good life available at the end of the North American rainbow?

Again, looking inward a bit, was I so immersed in the good life myself that I resented anyone else trying to get close to it, let alone get a piece of it? Had I bought into the new culture so effectively that I didn't care about family ties any longer?

As before, no clear answers emerged.

I haven't heard again from Bombay or Chicago since that phone call. Maybe my relatives are so miffed that they will never talk to me again. Or maybe they're hatching some other convoluted scheme to get my cousin into the United States. I don't know and I don't *want* to know.

But each time I hear of people found in transatlantic containers, rusty freighters disgorging illegal immigrants in New York and San Francisco and midnight crossings on unfrequented stretches of land borders, I wonder if that damned phone won't ring again.

And then what will I do?

Politics of survival

> So many people seem unable to distinguish between immigrants and refugees, between those entering Canada legally and asylum-seekers fleeing oppression.

As technology shrinks the world into a global village of easy communications and mobility, some of the repercussions are turning out to be unpleasant, if not ugly.

All over the world, those docile, teeming masses who have long been the fodder for political manipulation and upheavals are starting a revolution that is posing unique threats to governments.

Thanks to the rapid dissemination and availability of information, as well as to increased education, underprivileged people in the developing world are responding to their basic desire for survival with their feet.

By foot, by bus, by train, by sea and in planes, they're on the move.

Now migration for economic reasons is nothing new – it has populated Canada, the United States, Australia and Argentina, among other nations of immigrants.

What is new is the scale of present-day migrations – a

scale so staggering that a recent report by the United Nations Population Fund described it as an "uncontrollable tide" that could become the "human crisis of our age."

Today, according to the UN report, there are 100 million international migrants – which means that 2 per cent of the world's people live in countries they were not born in. And only 17 million of these immigrants are refugees.

Even so, the report points out that the search for a better life is generally confined to the country or continent of the migrants' birth and that over-all, poor developing countries shelter far more refugees than rich developed nations.

And it dwells at some length on the problems such internal migrations are creating in already crowded urban centres.

But the report also warns that the pattern of migration is international too, with large numbers of poor migrants, legal and illegal, seeking entry into rich countries.

The report couldn't have come at a worse time for the rich industrialized countries where new immigrants and refugees, despite their relatively small numbers, are already viewed as a threat.

Whether it's Japan or Germany, Canada or France, the doors are closing or are under pressure to close.

Phrases like Fortress Europe or Japan for Japanese are becoming more common, with inevitable imitations in North America.

Sadly, while it is easy to understand why they exist, I find it impossible to appreciate such attitudes.

I can certainly understand the desperation of the people affected by the tenacious recession in the Western world – a desperation almost akin to that of the impoverished people seeking entry.

I can also understand how those shiploads of illegal migrants from China are clouding the issues.

But somewhere, I feel, those who hold such attitudes

aren't paying enough attention to a crucial aspect of this debate – the facts, particularly as they apply to Canada.

The simple truth is that for every study that questions the need for immigrants, there are at least two that confirm their historical and continuing importance in Canada's evolution.

Further, people who disdain or devalue the contribution of immigrants are guilty of forgetting that they too or their ancestors were once immigrants.

They also seem unable to distinguish between immigrants and refugees, between people entering Canada legally and by invitation and asylum-seekers fleeing oppression. In terms of public policy, immigrants are a social and economic issue, refugees a humanitarian and political one.

Granted, there is a slight crossover when people seek refugee status on false premises or when refugees acquire legal status. But this should not submerge the basic distinction.

Those who fall prey to alarmist warnings about uncontrollable tides of migrants are also likely to become victims of associative prejudices.

From getting worried over immigration levels to accepting only certain kinds of immigrants is but a small step. And it's another tiny step to developing a dislike for humanitarian and other aid programs.

Ironically, these associative prejudices can often be a hurdle for solutions internationally that may make the prejudices redundant locally.

For example, according to the UN report, one of the ways to stem migrations is a greater investment by national and foreign governments in population-control and economic-development strategies.

Take away some of the conditions that force people to migrate and they'll stay home.

In some cases, it also means that talented and highly educated people won't leave, thereby stemming a brain drain that hurts the migrant's home country.

Immigration, with all its modern-day implications, will always be a touchy issue, but a little openness in the debate can stop it from becoming a divisive one.

Studies and statistics aside, let's not forget that migrants leave home with as much pain as they undergo in finding a new one.

Once there, circumstances permitting, it almost seems inhuman to deny others the same opportunity.

Dreams, just dreams

Wiping tears from his blue eyes and smiling bravely, Clive Bird offered a goodbye handshake with these halting words:

"Ca-na-da. I can't believe you're going to Canada.

"You've escaped, man – really escaped!"

Standing outside the Birds' modest railway bungalow in the central Indian city of Jabalpur on that muggy September day 20 years ago, I couldn't help shedding some tears myself.

> Being born in a particular place doesn't always make you feel you belong. Neither does moving to a new place. Sometimes, you're a foreigner wherever you are – at home or abroad.

I wasn't crying at the thought of leaving behind the country I was born in – indeed my impending departure was a source of immense excitement for me.

No, the tears were for my anglo-Indian friend, who had been a classmate for only a year in my life. But it had been a special year, 1963, when I was 13, wearing drainpipe pants with stitched creases, pointed shoes with raised Cuban heels, listening to Western pop music and acquiring more than a fraternal interest in girls.

And it was Clive, two years older, already worldly and streetwise, who had led (some, like the aunt I boarded with, would say misled) me through that first year of teenhood.

Outwardly, there was little in common between us. He belonged to a family made dysfunctional by the independence of India. An English family that had married some Indian blood in the hoary past, forever condemning it to a trapped twilight of incomplete belonging.

When the British left India, the few perks the anglo-Indians had enjoyed dwindled, helped partly by their own reluctance to assimilate with other Indians and partly because of the suspicion and contempt they ran up against.

Too poor to go "home" to England – most were in lower middle-class positions in the railways and bureaucracy – the anglo-Indians formed a cultural ghetto of imagined Westernisms, wearing hats on Sundays, eating curry with knives and forks, joking about "those bloody Indians" and holding Saturday-night dances where local musicians in Beatle haircuts belted out Top 40 hits.

I belonged to a family of Sindhi-speaking Hindus dislocated by the independence and partition of India, forced to flee the newly created country of Pakistan so they could remain Indian – but as refugees in their own country.

Although I had been born in Jabalpur, my parents had moved on when I was barely a year old. But I had returned in 1963 to stay with relatives and go to a school whose medium of instruction was English, something that wasn't available in the small town some 200 kilometres away to which my parents had moved in the forced nomadic existence familiar to all displaced persons.

At the St. Aloysius School, a Catholic-run secular school for boys, I found myself to be as much a minority as Clive, somewhat out of sync with the city we had been born in and not particularly at ease with the local language, Hindi.

It was inevitable that we be drawn to each other, two "foreigners" in their own land, a two-boy foil to our classmates and stern Catholic priests eager to cane our palms for the slightest of offences.

Besides, Clive knew a lot of anglo-Indian girls, friendly souls willing to teach you an old-fashioned foxtrot or the latest version of the twist or jive. Girls willing to play Spin the Bottle and Kiss in the Dark. Girls very different from the majority, who weren't allowed to talk to boys, let alone touch them. Girls who weren't allowed to play any games or go to any dances.

Not that it was as easy for me to go dances or to our female friends' homes as it was for Clive. It just wasn't part of my family's culture.

So, with some "guidance" from Clive, I had to invent all manner of events that would keep me away from home. Nonexistent after-school tutoring, punishments, class picnics or visits to museums or movies, and, the most frequent excuse, helping Clive with the homework he never did.

It wasn't that he was a slow learner. Just reluctant to let the joys of teenhood interfere with the demands of school. Six hard shots on his palms were so much easier to take than spending time writing a six-page essay. A truant afternoon splashing in or lolling beside the Narmada River a few kilometres out of town (we both had bicycles) was so much more fun, forged letter of absence from parents notwithstanding, than mathematics and moral science.

But heady as those days were, they weren't meant to last. The more time we spent together, the more mischief we got into, to the point where my aunt sent me packing to my embarrassed parents at the end of the school year.

I was lucky enough to pass that year. I can't remember if Clive was. And over the next nine years, as my parents moved to other cities and I moved through other schools and universities, I lost all contact with Clive. Such is the tenuous bonding among boys at that age.

But now, in 1973, living in the central Indian city of Nagpur, on the verge of immigration to Canada and marriage, an irrational impulse had brought me the bumpy 250 kilometres to Jabalpur – and Clive.

The encounter was short and sad. My boyhood pal's sandy hair was already receding. A second child was on the way. He was working in a dead-end job. He was even too embarrassed to invite me in.

We chatted a few minutes in the front yard and shook hands.

Wiping my own tears, I got back on my borrowed Lambretta scooter and the road – the last stop on a journey that would bring me a few days later to Canada and a new life.

A lot of water has flowed down that river Clive and I once splashed about in. And two turbulent decades have slipped by since my "escape" – years during which I have often taken heart from those boyhood memories of divided loyalties and identity traps.

So what if I sometimes feel – or am made to feel – I'm still a foreigner in Canada? The same has happened on my 10 or so visits to India.

And it's not at all different from how I felt – or was made to feel – in that school in my birth town.

Clive, wherever you are, there is really no escape.

Just survival.

Family? What family?

Do you have family here?

This is a question that I'm often asked, particularly when family-related issues are in the news.

Predictably, the question arose again last week after Statistics Canada released its latest 10-year portrait of the Canadian family.

> What binds people together and how does migration affect family and other ties? The answers are not always easy or readily available. To find any, you have to delve into the past.

And predictably, I found myself fumbling for an answer. Because, when it comes to defining family, all I usually come up with are more questions.

Does family mean one's blood relatives? Spouse and kids? Lovers? Friends? What binds people together and how does migration – forced or voluntary – affect these ties?

Groping for answers, I find myself looking inward and backward to the early 1950s, when fate brought five refugee families to Cuttack, a bustling provincial town surrounded by emerald rice fields in the eastern state of Orissa in India.

Abichandani the chemist, Vacchani the agronomist, Butany the botanist, Relwani the farm manager and my father the plant pathologist arrived at the Central Rice Research Institute among the paddies in suburban Cuttack sharing a bit more than a degree in agricultural science.

They were all Sindhi-speaking Hindus uprooted from their homeland, Sind, after it became part of the new Islamic nation of Pakistan created by the partition of India in 1947.

Choosing to remain Indian, they crossed the border into what was still India, against the backdrop of one of history's most bitter and violent ethnolinguistic and religious conflicts.

Fleeing bloodshed that cost hundreds of thousands of innocent Hindu, Muslim and Sikh lives and surviving squalid refugee camps, these five Sindhis found themselves in a linguistic state where almost everyone spoke only Oriya.

Everyone except their fellow researchers, who hailed from the subcontinent's other linguistic states and regions – Punjab, Bengal, Madras, Mysore and Kashmir, among others.

The only language staff at the institute had in common was English. And since many knew they would eventually be transferred or promoted to one of several other such institutes across the country, they felt it was important to educate their children in English instead of the language of the state.

Now this might have been a coincidence or an act driven by some subconscious ethnic memory, but of the 30 or so kids who were enrolled in Stewart School, an English-language institution, 18 belonged to the displaced Sindhis.

And even though the Sindhi families were in the forefront of such activities as staff picnics and celebrations, the 18 Sindhi kids always seemed to have a lot of time for each other. The kids, who ranged in age from 5 to 13, interacted on many levels.

In the baby-blue Bedford bus that took the kids to school 10 miles away, the Sindhis formed a strong majority able to exercise its will over such important matters as window seats.

In the school, which consisted of boarders from across the country and day scholars of various cultural and linguistic backgrounds, the minority Sindhis soon became a well-oiled machine able to hold its own in playground fights and protect such vital interests as (forbidden) fishing spots around a large pond that happened to occupy a corner of the sprawling campus.

They also proved to be wily traders in school, outdoing their fellow "farm bulls" in such matters as toy and gadget swaps. (The boarders called kids from the research institute "farm bulls" because of its rural, therefore primitive, location; this merely earned them the nickname "town rats" in return.)

Back in their 250-acre rural habitat, the Sindhi farm bulls seemed to spend most of their time in each other's homes, studying, eating and sleeping in whichever one they happened to find themselves in.

Without realizing it, these 18 kids, who spent only five years together before being dispersed by circumstance, became an extended family of 10 parents, eight sisters and 10 brothers.

In hindsight, this unconscious clannishness is intriguing.

None of the kids or their parents had known each other before they arrived in the multicultural and multilingual environment of the institute. Almost all the kids had been born after the traumatic events of 1947. None of them had acquired any deep sense of their language or heritage – they could only speak Sindhi as a mother tongue and had no opportunity to learn how to read and write it. And although they spoke Sindhi with their parents, English was the language they used with each other.

It was also English – specifically proficiency in the language – that ultimately broke up the newfound clan as the 18 grew up to pursue careers in science, psychology, law, medicine, engineering, defence, business, advertising and journalism.

And of the 18, 14 immigrated to such cities as Kuala Lumpur, Los Angeles, Tripoli, Detroit, Toronto, Montreal, Calgary and Saskatoon – all of their own volition, unlike their parents, who had been forced into migration.

Today, some of the parents of this scattered family are dead. Others remain in India, while one couple lives in Toronto.

Almost all the 18 kids are parents themselves, their children native-born citizens of new countries like Canada, more adept at English and French than Sindhi and somewhat puzzled by the long-lost "uncles" and "aunts" who sometimes phone or appear at their door.

The new kids' confusion is endearing. They have many years to go before the bonds they're forming now will assert themselves. For now, they're free of the tangles of migration – the death of old ties and the birth of new.

Clinging then to a dangling thread or two, answering that question about family here, all I can come up with is: sure, I do.

But then I can't help adding: I have family elsewhere, too.

ASHOK CHANDWANI

A matter of respect

Even as Canadian, American and Asian leaders were gathering in Seattle for the much-heralded dawning of the Age of the Pacific, Vancouver, a principal player on this circuit, was demonstrating in a small but disturbing way the troubled road to tolerance and friendship.

> When it comes to flaunting ignorance, some Legionnaires in British Columbia aren't alone at all. Prejudiced attitudes based on cultural stereotypes can be found all over the country.

If the underlying assumption of forging closer links among the Pacific Rim nations is mutual tolerance and respect for the diverse peoples and nations involved, then it's a message that has yet to filter down to the members of the Newton Branch 175 of the Canadian Legion in suburban Vancouver.

Despite widespread condemnation from political and community leaders as well as from enlightened Legionnaires from other branches, Newton veterans have dug in their heels on the issue of admitting turbanned Sikh veterans into their Legion Hall.

We have nothing against Sikhs, the Newton vets say, just

their turbans. Our club rules on headgear demand that members and visitors to the main hall uncover their heads out of respect to the Queen.

It doesn't seem to matter to the Newtonians that the same Queen has in the past, and still does, receive Sikhs wearing the turbans that are an integral part of their religion. Or that other Legion branches do not have such exclusionary rules.

It also doesn't seem to matter to these vets that millions of Sikhs, Hindus, Muslims and Christians from India fought in the same wars the Canadian vets did and that thousands of their graves dot the same killing fields in Europe, Africa and Asia.

It also doesn't seem to matter that many of the "other" veterans earned the same Victoria and other crosses for their courage and valor on the battlefield.

Indeed, one of the five Sikh veterans who was denied entry into the Newton Legion Hall was a 92-year-old who was awarded the Order of the British Empire.

The irony in the Newton branch's rules, which were reaffirmed by a vote last week, is that they ignore the reason why Sikhs wear turbans in their gurudwaras and in public – as a mark of respect.

As the World Sikh Organization of Canada pointed out in a statement last week, "the very thing we wear to show our respect is now considered disrespectful."

The matter is not closed yet. Some of the barred Sikhs are considering filing a complaint with the province's human rights commission. And the Legion's dominion executive council meets in Ottawa later this month, where a national directive on headgear rules in branches may be discussed.

Whatever the outcome, there's little doubt that this incident is yet another in a chain linked to the lingering prejudice against not just Sikhs but all other Asians in B.C., notably people of Chinese origin.

Any policy of closer economic and social ties with the

Pacific Rim will have to take into account this prejudice, even though it's no longer on the scale faced by the Chinese and Sikhs when they first arrived in B.C. more than a century ago.

It's always tempting to explain the prejudice as a Western problem. As a friend who attended the Remembrance Day parade in Montreal remarked:

"No one gets excited about turbans or other headgear here. Why, this year, for the first time, I saw some Mounties wearing blue UN berets in the parade. They had clearly put them on instead of their traditional hats because they were proud of their UN service.

"But nobody jumped up and down because the Mounties had changed their headgear, the way some people did out West when that Sikh Mountie joined the force."

Yes, it's tempting to blame the West when it comes to overt prejudice against Sikhs and their turbans. But the West does not have a monopoly on prejudice. Even as the B.C. case unfolded last week, a Montreal professor offered disquieting testimony at an unprecedented public hearing by the Quebec Human Rights Commission into violence and discrimination against homosexuals.

Frank Remiggi, a geography professor at the Université du Québec à Montréal, told the commission that recent field visits he conducted with police-technology students were "the most painful and difficult in my 15-year career."

"Queers, fifis, homos" were some of the epithets used by the students when they toured east-end gay districts, Remiggi testified.

Remiggi made it clear that not all the students held such views. But enough held them to become police officers because 80 per cent of his groups always do.

The future police officers' prejudice extended to other minorities, too. One student said she could never touch a black person because she believed they were dirty.

Another student said there were two neighborhoods she could never live in: "Little Burgundy (a black area) and the Jewish Quarter (Outremont). For a white francophone like me, these areas seem repulsive."

Another student felt that the distinctive dress and skullcaps worn by Hasidic Jews should be banned. And one, after a visit to Little Italy, said, "It smells Italian."

Remiggi said he was concerned about "rising intolerance" among young Quebecers and urged compulsory social-differentiation courses, especially for prospective police cadets.

So there you have it. On one side a group of aging veterans clinging to a bylaw that entrenches prejudice. On the other side, groups of young Quebecers clinging to racist stereotyping. Neither group represents a majority or, for that matter, a geographical view.

But the fact that such ignorance persists on both sides of the country should concern all of us.

Riches, here I come

Like the P.G. Wodehouse character Ukridge, I spend many pleasurable hours dreaming up and chasing money-making schemes.

Like Ukridge, I always seem to get pre-empted by circumstance. Or if the project does make a shaky start, it crashes almost instantly.

But mysteriously, another scheme appears from nowhere and sucks me right in.

> With so much gold supposedly paving the streets in the new country, it's not surprising that newcomers are always cooking up money-making schemes. Not everyone strikes it rich, though.

When I'm not succumbing to these schemes, I find myself wondering about my weakness for them.

Could it be the childhood influence of an enterprising uncle with a similar disposition?

Could it be the I'll-show-you-what-I-can-do effect of studying in schools where everyone else was from a comfortable middle-class or rich background?

Sometimes I feel I have fallen prey to the Immigrant Dream – the streets are paved with so much gold and so many strike it rich so often that it's downright traitorous of me to not even try.

It's not financial desperation that motivates me. Like that uncle of mine, I do have a comfortable job. Maybe it's the exhilaration of it all, tinged with misplaced glamor.

I remember vividly talk by and about my uncle when I was a child. He worked for a major oil refinery in Bombay. In the early days of his working life – he's retired now – his job was to supervise the transfer of oil from super-tankers.

Naturally, this involved extensive and repeated contact with all kinds of foreign captains and sailors. And naturally, my uncle always seemed to be able to procure foreign watches, fountain pens, electronic gadgets, cigarettes and fancy liqueurs that were either banned or prohibitively expensive in the then-protectionist Indian market.

You'd think my uncle would have made a lot of money from all this. But you'd be wrong. My uncle never seemed to know how much to charge for his goods, many of them being gifts. So he'd give a few away, too. At other times, when he did set a price, he never seemed to be able to collect all the money right away. As for the liqueurs, well, he often found other pleasant uses for them.

Later, when he found himself moved to the head office, cut off from his little sideline, my uncle decided he would become part-owner of a truck. A partner would supervise its operation, hire a couple of drivers, fill it with cargo and the money would just roll in.

In their eagerness, my uncle and his buddy forgot that it takes a lot more to keep a truck rolling in India than cargo and a driver. Things like inter-city and inter-state permits and greasing not just palms but occasionally the truck itself.

What can I tell you? Finally, it was the truck that got my uncle out of that hole by falling into one – a huge one caused by heavy monsoon rains.

My uncle never told me so, but I'm certain it was a modest insurance payoff that he then invested in a corner pharmacy. By then it was time for him to retire from the oil company. So it really didn't qualify as a get-rich scheme. More like a make-work project.

The last I heard – I had left India in search of my own fortune – it isn't doing too badly. Although, when I do take the time to phone, my uncle's always stepped out of the pharmacy for a cup of tea and at home he's invariably taking a nap.

Ah, contentment, you say. That's what it's all about, isn't it? I couldn't agree more, but, alas, I'm still miles from it.

I do recognize, however, that somewhere at the back of my irrational mind, it must be the prospect of some form of contentment or bliss that makes me vulnerable to my own schemes.

Take the business of sausages. Being fond of and having tasted so many kinds, I must have spent at least a year discussing the Great Sausage Bazaar with a close friend of mine.

Our plan was to start not just one but a whole chain of shops that would serve as many kinds of sausages as there are countries that make them. And we'd serve them with breads, pickles and condiments that would be authentic to the particular country or cuisine. Not to mention appropriate beers from those countries.

By the time we had thoroughly fleshed out the idea, sampling far too many of those appropriate beers, someone else beat us to it. Not with quite the range and diversity we had sought, but enough to pre-empt the market.

Hindsight also revealed a fatal flaw in our plan – we didn't have the money to make money.

Which is why all my bones rattled in unison when a friend came up with a fax-based scheme to make a mere million or two.

All we would need was a third partner, preferably in the shipping business and with access to a fax machine.

The key to the scheme was my friend's connections in

Bangladesh and that country's need for cement. Apparently, the country was accepting bids for a 600,000-tonne contract for cement. If we could come up with a competitive quote, he was in a position to ensure its acceptance through a highly placed relative. Our money would come by adding on a finder's price in the middle.

You bet we'll make money, a shipping friend told me. Leave it to her, she added. And for a while, it seemed she was right.

There were unexpected hurdles, of course. For political reasons, Bangladesh did not want cement from neighboring India. An Islamic country would be preferable. A socialist or communist country would be fine, too.

The Bangladesh port of Chittagong was too shallow for a big freighter, so a ferry barge would have to be arranged. And they wanted the cement in polythene bags of a certain strength, not open stock.

After weeks of faxing and phone calls to everywhere from Brazil to Hong Kong to Dubai, my shipping friend found a supplier in Romania. The price was right, a little less than $40 a tonne for open stock. It would cost us a bit to have the stuff bagged on the high seas as it sailed to Bangladesh, but even so we would be able to quote a price that would put about $3 a tonne in our pockets.

The ship was arranged. The barge was arranged. The cement was arranged. Nothing could go wrong, but it did.

Ceausescu was deposed in Romania, the fax lines went dead and the country was plunged into turmoil from which it would emerge weeks after the deadline on our contract.

Easy come, easy go.

ASHOK CHANDWANI

Looking for bonds

Many years ago, a friend told me one of those stories only minorities seem to tell – a potentially tense encounter with the mainstream unexpectedly defused by wit and common sense.

While receiving her Canadian citizenship, this friend found herself balking at swearing allegiance to the Queen.

She explained to the citizenship judge that she was a republican from India, a country that had severed legal ties with the monarchy three years after independence in 1947. As such she wasn't really all that fond of the Queen.

"Madam," the judge replied soothingly, "I am a French Canadian and I don't like the Queen that much either. But that's the law and you have to abide by it."

I thought of the judge's disarming candor and its implications last week when the Superior Council of Education released a 132-page government-commissioned report on

> Pluralism in Quebec faces challenges as distinct as the province itself. History, linguistic pride and legitimate fears of assimilation all have an impact on francophones.

ethnic Quebecers and their role in schools and society.

The report attempts to come to terms with some of the issues faced by my friend and her judge more than 20 years ago and indeed by all nations that welcome immigrants.

In leaving one country and culture for another, what are the expectations that come into play, not just for the newcomer but also for the established groups? And how does one ease the almost guaranteed clash between the cultural and historical baggage of the migrant and that of the host?

In bringing my friend onside, the judge invoked the basic bond among minorities – we're both in this together and we have to cope with it jointly.

But this form of co-opting, which occurs frequently at the personal level, is unfortunately too facile and impractical on a larger scale – particularly in Quebec, where the majority population is itself a minority in the larger entities of Canada and North America.

The challenges posed by pluralism in Quebec therefore are as distinct as the province itself, colored by the history, linguistic pride and legitimate fears of assimilation in the francophone majority. Factor in the urban-rural divide and the concentration of newcomers in the Montreal region, and the equation gets even more complex.

It's no wonder that all kinds of committees and interest groups are involved in finding an equitable and acceptable solution – a social contract that anchors certain established values, but doesn't close its mind to new ones.

The problem – or should I say conflict – lies in defining the values and ways of enshrining them.

No one can argue with the council when it suggests that the integration of immigrants into their new society calls for a two-pronged approach that starts in school. Newcomers have to be introduced to the existing value system in Quebec, preferably early in life at the school level. But the same value system, the council says, also has to be taught to francophone Quebecers – even in places like Lac St. Jean, where there are hardly any immigrants.

Old-stock francophones need to understand the rules of the game in a Quebec that is being altered by new waves of immigration, council chairman Robert Bisaillon told reporters last week.

However, in trying to define this value system, the report bases its vision of Quebec's "common public culture" on a few puzzling, if not questionable, assumptions.

The council is on solid ice when it notes that this common public culture is based on French as the official language, respect for the anglophone minority, a legal system derived from British common law and French civil law, a charter of rights, a parliamentary democracy and an economic system that includes private enterprise and state-operated companies.

But the ice starts to thin when the report says such a common culture draws inspiration from a Judeo-Christian tradition based on French, British, American and aboriginal sources. Aboriginal?

And further, it says immigrant kids must do more than learn French and graduate from high school if they want to fit into the French mainstream of Quebec society.

These statements raise disturbing questions. What precisely does the council expect from immigrants? By its own admission, nine out of 10 children born outside Canada attend French schools; over-all, they perform as well and often better scholastically than do francophones. So where exactly are they falling short and how or where are they out of step with the common public culture?

While Bisaillon made it clear that the public culture described is not code for white francophone values, he also expressed concern about the uneven distribution of immigrant kids in the school system. In three-quarters of Montreal's French schools, fewer than one-third of students were born outside the country. But in 26 of Montreal Island's 304 schools, more than half of the students were born outside Canada.

For Bisaillon, this concentration poses a potential problem, mainly because it gives old-stock francophones and new Quebecers little chance to get to know one another. But in suggesting school boards redraw boundaries that determine where kids go to school, he's back on the same thin ice as with the Judeo-Christian business.

I find both issues in some conflict with the content and spirit of Quebec's and Canada's charters of rights. Both seem to involve some form of unreasonable regulation as well as exclusion.

When you talk of redrawing school boundaries, you're trying to limit the freedom of people to live where they choose and dictating which people they interact with in some absurd frame of optimum contact.

And when you talk of a Judeo-Christian ethic, you're sidelining the pluralist and secular foundations of any enlightened public culture.

While it's inevitable that any public culture will be heavily influenced by the religious traditions of its dominant population, a secular and increasingly pluralist society is obliged to accommodate the traditions of its minorities as long as they're consistent with the laws of the land.

To do otherwise is to open the door to dated forms of cultural chauvinism and assimilation. And the possibility that any such moves would be fuelled by fears of assimilation by wider entities only injects an element of pathetic irony into the entire debate.

As my friend and the judge unwittingly demonstrated, it's preferable to confront and accommodate our collective fears than to quarrel or adopt regressive positions about them, no matter which ethnic group we belong to.

'Tis the season

Rev. Lawrence Fernandes walked slowly down a line of teenagers tearing their lungs out in a cavernous, stone-walled room and bent an ear barely a foot from my mouth.

> In December, not a day goes by without someone asking whether Christmas is celebrated in India or wishing me a merry one in tones that assume it's my holiday, too.

He listened for a minute, signalled for a break in the singing and delivered a tactful verdict:

"I think you'd be better off in the drama and debating clubs."

But for the twinkle in his steely eyes, I was ready to burst into tears that fateful night 25 years ago as I confronted my stillborn musical career.

There goes a chance to go Christmas caroling, I gulped, and the chance to join the Choral Society, of which "Father Lawrie" was the conductor and guiding force. At the time, Father Lawrie was also the head of the English department at the St. Francis de Sales College, which sits atop the Seminary Hills in the central Indian city of Nagpur.

Later he would become the principal and hire me as a lecturer in the department he headed, but on that day of the singing audition I was still a second-year college student.

Curiously, it was on that December night, in the prewar mansion that housed the college administration and quarters for senior priests, that the ground was laid for a friendship that survives to this day.

It is an unlikely alliance. He's an erudite Catholic priest and educator and I'm an agnostic born of Hindu parents.

Over the years, with me as his student, teaching colleague and friend from abroad, we've found time to engage in spirited debate and we've found time to savor a warm cognac or two.

Not once during these years, whether he's been discussing such issues as Marxism, abortion, secularism and religious fundamentalism or officiating at my engagement to a Catholic, has Father Lawrie brought up the question of my religious affiliations – or lack of them.

This is not because Father Lawrie is not a devout Christian – indeed, his own faith has always been remarkably steadfast. Nor is he one of those politicized people's priests that developing and Third World countries have spawned in recent years.

No, Father Lawrie, in a truly modern and enlightened way, respects the right of others to their own views and religions without losing faith in his own.

So what's so special about that, you might ask. Isn't that how it should be? Isn't that how it is in any civilized society? Isn't that how it is here?

Not quite.

It's easy to believe in secularism, but extremely challenging to practise it. No matter how up-to-date the laws and constitutional guarantees, the dominant religion of a society is bound to affect its social and political attitudes.

For Father Lawrie and 17 million Christians in predominantly Hindu India, this means that official ceremonies and practices bear a Hindu stamp, even though the secular constitution clearly separates state from religion.

And while the same constitution ensures that Christmas is a national holiday along with the principal days of all other major religions in the country, as a minority he is constantly – and inevitably – bombarded with the attitudes and expectations of more than 700 million Hindus.

The point here is that while laws are indispensable, it is ultimately up to people to propagate their spirit. And it is how people do so that determines the levels of tolerance in a society.

I bring up my friend Father Lawrie because every time Christmas looms I am reminded of his minority status in India even as I am made conscious of my own in Canada.

For not a day goes by without someone asking me if Christmas is celebrated in India or, as the office parties begin, wishing me a merry one in tones that unquestioningly assume that I celebrate the holiday, too.

I can never decide which situation to be more frustrated over: the suggestion that people in India, and therefore someone from there, would be ignorant about Christmas; or the presumption that I'm running around shopping for a turkey and gifts like everyone else.

So I find myself explaining politely that some Christians in India trace their heritage to the first century AD, that there are thousands of Christian-run schools in that country and that I was married to a Catholic. So, yes, I know all about Christmas.

Then, I also find myself challenging some people's attitudes about Christmas, the basic one being that everybody celebrates it and if they don't they should.

OK, no one quite says it that bluntly, but the message certainly comes in loud and clear to anyone who's not Christian.

It comes through in the pre-Christmas events and activities in schoolrooms; in the huge Christmas trees in government and public institutions; in all the commercial hoopla in shops and advertising and in the sudden flurry of charity fundraising.

Even if it isn't intended that way, it's a message of exclusion that can and does disturb minorities, particularly children – as many non-Christian parents can testify.

So what's the answer? Stop celebrating Christmas in public? That would be absurd.

No, the answer is really quite simple and, yes, quite Christian!

A bit more tolerance and a little more love for our fellow beings. And more participation in each other's festivals and ceremonies, religious or secular.

I'm ready for my share. I might have flunked that carol audition 25 years ago, but it doesn't mean I've lost all hope.

Thanks, Miss Francis

Sorting through boxes of papers and documents – a futile urge I succumb to once a year – I found a tattered pink card that I had forgotten I owned.

Pink was the color of honor at the St. Joseph's Convent in Chandernagore, a former French colony on the Hoogly River, some 20 kilometres upstream from Calcutta.

> I can't swear to it, but I'm pretty sure it was my teacher's sparkling eyes, her sunny smile and her spontaneous hugs that kept me motivated in elementary school.

And that's what it says on this card – Card of Honor, First Class.

Other information in neat handwriting on both sides of the card reveals that it was awarded to me, in Standard IV, and that my conduct was good. Good marks 41, bad marks 11, total marks 74 per cent, distinctions in English Text, Sentence-Building, History, Geography and Arithmetic.

All these achievements when I was only 8 years old – a year before they threw all the boys out of the girls' school – had faded completely from memory, only to be rekindled by the palm-sized card.

But try as I might, all I can dig out of my subconscious

of my two years in that school are blurry images of little girls in white blouses and navy-blue tunics, dark, long hair in twin plaits secured with white ribbons. And I remember one of the teachers, a Miss Francis.

The girls I can understand, but why I remember Miss Francis is intriguing, given that most of the teachers at the convent were nuns. Perhaps it was because she was a "civilian" that Miss Francis sticks in my mind.

Or perhaps it was the running battles my father had with the nuns that have banished them from my memory. The nuns, you see, were suspected by my father of adding a little extra on the price of school notebooks and things in the guise of school crests and logos. It didn't matter that the extra pennies probably went toward school expenses. My father was convinced the money went into church coffers.

"Why should I pay extra for a notebook with a school crest," my father would rage, "when the same notebook of the same size with the same rules is cheaper elsewhere?"

"I'm sending you to the school for education, not to raise money for the church."

I can't remember whether he won all the battles. But I do remember my embarrassment as I went to school with a notebook unadorned with the school crest or a textbook protected with plain brown paper instead of a sheet with the school name on it.

The embarrassment, I recall, would turn into alienation from my classmates, whose parents dutifully obliged all the nuns' requests.

But countering that dimly remembered exclusion – a majority, even if it consists of kids, has a way of rubbing things in – is the warm and comforting presence of Miss Francis.

I can't swear to it, but I'm sure it was her sparkling eyes and sunny smile, framed by a fashionable bob cut, and her spontaneous hugs that kept me motivated.

For even at that young age, we children were always under pressure to compete for that pink card.

The heat came from the teachers, from parents and from other kids. But ultimately, it came from the class teacher, the one who taught most of the subjects, did the roll call and was doctor and nurse to everyone. It came from Miss Francis.

I bring up Miss Francis in the context of the intense discussions raging across the country these days on the nature and content of education and its role in helping us cope with a restructured economic order.

As companies lay off thousands of people and jobs die or disappear down the information highway, governments and assorted experts are focusing on a revamped education system to overcome the challenges ahead.

Some of the impetus for zeroing in on education is coming from the experience of such countries as South Korea, Taiwan, China and India. These are all countries that have made heavy investments in education, often at the expense of social programs.

For South Korea and Taiwan, the investments have paid off handsomely – their economies have taken off, bringing a vast improvement in living standards.

For China and India, which have massive populations and poverty, the yield is slower, but real. The developed sectors of both economies are booming, with India making impressive gains in such areas as software engineering.

The continuing poverty in both nations notwithstanding, it is the emphasis on higher education that is being credited for their success in fields traditionally dominated by the West.

Not surprisingly, therefore, much of the talk about educational reform these days revolves around the curriculum and methodology. There appears to be a growing feeling that we have to return to certain basics – the three Rs and a dose of old-fashioned discipline, if not a severe regimen in the classroom and at home.

The reformers also advocate a greater emphasis on science, mathematics and computer science as some kind of magic formula that will bring back the good times when most jobs were for life and the unemployment rate a low single digit.

But lost somewhere in this commendable debate about scholastic performance and a curriculum for the 21st century are the people who will be the key to achieving these miracles – the teachers.

Too often teachers are given a back seat in such discussions, even as they take a back seat in their students' minds as they grow up and scatter.

A few of them might deserve such a fate, but hundreds of thousands don't.

The role of teachers is paramount in the classroom, irrespective of the system of education, the ethnic background of the students and the economic status of their parents.

And it's not just how well-versed in their subjects teachers might be, it's also how they communicate and interact with their students. It's how they establish personal rapport through love and sensitive attention. A little pat here, a little understanding there.

It might sound like a thankless task, but it is not. The rewards are always there, always tangible, even if they're not immediate.

So as 1994 shifts into gear, I'd like to raise a grateful toast to all such teachers, the true unsung heroes of these troubled times.

And that includes you, Miss Francis, wherever you might be.

The rules of Rome

When you are at Rome, live in the Roman style; when you are elsewhere live as they live elsewhere.
— **Advice to St. Augustine from St. Ambrose in Ductor Dubitantium, a book written by Jeremy Taylor, an English divine, in 1660**

In a pluralist society like Canada, the laws or practices of other countries should not be used as an excuse to flout or undermine our own democratic traditions.

I have never read the English divine's writings. Nor do I know much about the life and times of St. Ambrose of Milan, the fiery fourth-century preacher and scholar who converted St. Augustine to Catholicism.

But according to assorted books of quotations and encyclopedias, Taylor and St. Ambrose are the two luminaries responsible for the proverb most of us know as "When in Rome, do as the Romans do."

It's a proverb – and local variations of it – that seems to crop up frequently these days in a context and pattern that is extremely disturbing.

Listen to Angelo Lebano, a veteran Cornwall, Ont., city

councillor and president of the local branch of the Royal Canadian Legion, who was forced to step down pending a disciplinary hearing from the provincial command after he made some comments about Sikh veterans and headdress.

"When in Rome you do as the Romans. If they (Sikhs) feel they have to wear their turbans, then let them go back to their country and do what they want to do."

Listen to Bill Doolan, president of the Legion's Branch 12 in Sydney, N.S., which has never allowed religious headdress. "Canada's becoming so ... deluded," Doolan told the press a couple of weeks ago.

"This makes me sick. I'm tired of bleeding-heart people making demands. When we go to Asia, we take our shoes off before we enter their houses."

And listen to Municipal Court Judge Richard Alary, who is facing a disciplinary hearing by the Quebec Judicial Council after he expelled Wafaa Mousslyne, a Muslim woman charged with shoplifting, from his courtroom in Longueuil because she was wearing a hijab, a religious headscarf worn by many women of the Islamic faith.

According to court transcripts, Alary, after expelling Mousslyne from his court, told her lawyer:

"When one goes to Rome, one lives like the Romans," adding, "If I go to Saudi Arabia, my wife wouldn't like it (because she has to follow Saudi custom)."

So, you say, there are misinformed and insensitive people all over, even among the judiciary and war veterans.

And didn't the top brass of the Royal Canadian Legion prohibit any branch from denying entry to veterans wearing religious headgear, be it turbans or yarmulkes, after that incident in Newton, B.C.?

Wasn't it the Quebec justice minister who promptly called for an inquiry into Alary's conduct, even as other jurists and human-rights activists were denouncing it?

Very true. And very commendable.

There's no doubt that officialdom is on the side of the angels on this issue. And indeed on other related issues such as the deplorable and ill-informed comments and positions judges have been guilty of in cases of sexual abuse.

Even the nation's top judge, Supreme Court Chief Justice Antonio Lamer, has stated publicly that racist or sexist behavior on the bench will not be tolerated.

Nevertheless, the message seems to be falling on deaf or stubborn ears, if the number of incidents is any guide.

Elements of this message – and perhaps views ignoring it – are bound to be aired during the next few days as the Federal Court of Canada begins hearing a lawsuit challenging the right of Mounties to wear turbans.

A motley group of RCMP veterans and their supporters has raised more than $100,000 to fund this challenge, backed by a 210,000-signature petition against Mounties wearing turbans.

Defending the three-year-old RCMP policy that allows Canada's only Sikh Mountie – Baltej Singh Dhillon, who's posted in Quesnel, B.C. – and any future ones to wear a turban will be Canada's attorney-general, solicitor-general and RCMP commanders. The Canadian Human Rights Commission, Alberta Civil Liberties Association, Alberta Inter-Religious Coalition, Sikh Society of Calgary and World Sikh Organization are other groups that are expected to play a role.

Resolution of the case is expected to highlight a fundamental issue: the constitutional right of minorities to be treated fairly and equally.

It's a simple issue really. Even though our constitution is not technically a secular one because of its reference to God, the federal and Quebec charters of rights specifically support a pluralist society in which the rights of all citizens are equal, irrespective of their race, gender, religion or ethnicity.

The issue is not the laws or customs of other countries or those that some new Canadians may have come from. Whatever Judge Alary's wife might experience in Saudi Arabia or Legion member Doolan might feel obliged to do when he visits Asia is totally irrelevant to what happens here.

In Canada it is *our* laws that are the issue and *our* constitutional and social commitment to a pluralist, liberal democracy.

And as far as new Canadians are concerned, now that they're in Rome, they expect to be treated like other Romans.

ASHOK CHANDWANI

A two-edged sword

A reader phoned last week to tell me about a forthcoming event – some sort of India celebration – and suggested I write about it because we were both of Indian heritage.

Simple as that. We happened to have been born in the same country, so it seemed a completely natural and logical reason to her for my writing about this event.

> In preserving certain cultural practices, minorities have to dump other dubious ones. Like voting for someone because he or she is from the same ethnic background.

It took several minutes of patient discussion – shouting doesn't help in such situations – to convince my caller that a common ethnic heritage is rarely a good reason to write about an event. That there has to be some news value to it or some implication that makes it interesting to a larger audience.

The caller, like many others over the years, hung up informed, if not satisfied, by my response. But, as in the case of previous callers, she left me wondering again about ethnicity and its public role.

I have absolutely no quarrel with ethnicity in its private existence. For an overwhelming majority of us, ethnicity spells culture spells identity. In a free society, how one chooses to manifest this culture and identity in private remains the legal and moral preserve of the individual.

The conflict arises when the expression of ethnicity spills over into the public culture, which, for obvious historical reasons, is heavily influenced by the ethnicity of the majority.

Typically, the majority does not see its culture as an expression of ethnicity, but rather as a universal norm that other ethnic groups are expected to conform to.

The fact that a culture begins with a particular group, expands with its growth and entrenches itself as the group gains social, economic and political power is often lost in any discussion of cultural interaction.

Forgotten also is the fact that any viable culture changes with the times, its latest manifestation being the result of decades, even centuries, of evolution.

This might explain why a majority is generally suspicious of and resistant to minorities. And why minorities try doubly hard to reduce this suspicion and be accepted.

Inevitably, the two opposing forces achieve a fragile reconciliation, sometimes through enlightenment, sometimes through legislation.

This, to my mind, is the kind of uneasy social truce that has gradually taken hold in Canada in the last 20 years as the entrenched English and French majorities have responded to the rapid increase in the number and diversity of minority groups.

This truce, which is bound to be elevated to the status of a tradition one day, hasn't been achieved without pain. A cursory glance at the cases human rights commissions have handled over the years will tell you that.

But with the pain have come real gains. Today, Canada

can legitimately boast of being not just a multicultural society, but a reasonably peaceful one. It can boast of a Parliament of unprecedented diversity. And it can rightly claim that it hasn't trampled unduly on ethnic pride in the process, leaving minorities room to preserve and contribute some of their cultural traditions.

The challenge for minorities then is to respect this hard-earned accommodation.

I am not suggesting some kind of gratitude-linked grovelling attitude. No, all I'm saying is that in preserving certain cultural practices, minorities have to dump other dubious ones. Like expecting favors from someone because of shared ethnicity. Or voting for someone because he or she is from the same ethnic background or speaks the same language.

This is the real social contract behind the cultural mosaic that is the new Canada. It's a contract that has brought minorities a lot of benefits and a lot of responsibilities. Put simply, freedom has to be a two-edged sword to be effective.

It may be unpalatable to many to hear the Bloc Québécois and the Reform Party bleat against multiculturalism. But they do have a right to their distorted views, as much as the rest of us have to denounce them.

By the same token, when someone from a minority group undermines this social contract, minorities should also be the first to denounce such a person.

I'm talking now of Jag Bhaduria, the India-born MP from Ontario who is under severe pressure to resign from the Commons amid controversy over threatening letters the former schoolteacher wrote to the Toronto school board in 1989 and over allegations that he falsified his academic credentials.

Bhaduria has already quit the Liberal party, on whose ticket he ran for election – a nomination he acquired by overwhelming support from newly signed-up Liberals of Indian heritage – but is stubbornly resisting calls to quit the Commons, too.

It has been clearly established that Bhaduria displayed poor judgment in his angry letters with their unpardonable references to woman-hater and murderer Marc Lépine. Shortly after his election last fall, he showed the same questionable judgment in appearing as a character witness for Kuldip Singh Samra, a man who had already admitted killing two men and wounding a third in a 1982 courtroom shooting.

Now, whether or not Bhaduria falsified his academic record, you'd think someone from the Indian community would have stepped forward by now and joined the growing public clamor for his resignation.

No one has, so let me be the first.

Quit, Mr. Bhaduria.

ASHOK CHANDWANI

Running spooked

Is immigrant becoming a dirty word?

These days, a stranger eavesdropping on Canada could easily be forgiven for thinking so.

Everywhere you turn, immigrants seem to be on people's minds and tongues. And, to a curious eavesdropper, leading rather strange lives.

While some are busy stealing our jobs, others are swelling welfare and jobless rolls.

> Anti-immigrant rhetoric, either pointed or oblique, forces people to flirt with questions they dread. Should I go back home? Can I go back? Is it still home? Where is home?

When they're not bringing a sixth or seventh child into the world, they're importing hordes of relatives, none of whom speak English or French.

When they're not buying mansions in West Vancouver, they're crowding 10 to a tiny room.

And they're becoming pesky about culture, refusing to learn new ways and customs and gobbling up public money to promote their own.

Why, some of them won't even pray to certain gods, choosing to stay with their own, even in the House of

Commons. Kidding aside, to use a phrase fast gaining common currency, what is this country coming to?

As an immigrant who mercifully won't be required to carry a plastic ID card because he became a citizen a long time ago, I can't help but respond with a bit of the same paranoia that appears to be guiding Reformers and a whole bunch of otherwise incompatible people.

It's a sorry and disturbingly unified message that's coming from these disparate voices, the latest being those of ex-Liberal Jean Allaire and his newly formed Action Démocratique party in Quebec, with their ill-defined and ill-conceived plan for written guarantees from immigrants that they will "settle, live and prosper in French."

It's a message that undermines the validity of my existence as a new Canadian – a "rebirth" that Statistics Canada tells me I share with the 16 per cent of our population that is foreign-born.

For the first time in my 21 years in Canada, words like demonization, dehumanization and marginalization are taking on chilling personal dimensions.

If there's one attitude all newcomers arrive with, it's a desire to blend as unobtrusively as possible in the new society.

This attitude is shared by traditional immigrants – those who voluntarily leave home for permanent residence in a new country – and displaced persons like refugees, who are involuntary immigrants, forced to flee their homes by forces beyond their control. If anything, that South American exile who dreams and writes of his home country is even more careful to adapt to local customs, driven as he or she is by sheer gratitude for having been given shelter.

As much as newcomers are keen to be accepted rather than tolerated (an important distinction), they also quickly sprout a second hide to withstand inevitable resentment and intolerance from those whose ancestors were also once first-generation immigrants.

Sometimes this second skin works in reverse, imprisoning newcomers in isolation and forcing them into mental or physical ghettos – or both.

But those of us who avoid or escape such ghettos – and we're a healthy majority of newcomers from a hundred different lands – use this extra skin as the necessary price of transplantation.

Put simply, we ain't dumb and we ain't presumptuous – we know that it takes time to be accepted and to plant new roots.

Lately, however, the stridency and persistence of anti-immigrant rhetoric, be it pointed or oblique, is taking its toll on such defences.

As my own second skin stretches thinner and thinner under this onslaught, I find myself flirting with questions most immigrants dread and deny. Should I go back home? Can I go back? Is it still home? Where, for god's sake, is home?

There are no real answers to these questions, of course. And any glimmer of an answer is ultimately personal and insignificant in the larger scheme of things.

One person fleeing Paki-go-home cries is not going to affect Canada's immigration policy or the attitudes of those who dislike it.

Besides, there are probably two new immigrants with even thicker skins out there, eager to replace each spooked fool running in the opposite direction.

I guess I'll just have to keep working at it.

Cause for despair

A cab driver, racing along snow-covered Montreal streets, pauses in an unprovoked midnight discourse about his business ventures to ask: "Excuse me, what is your 'nationality'?" Relieved by my response he carries on: "... and this Jewish guy screwed me. You know how these people are in business, eh?"

> What do we really think of each other? And are we as noble and sensitive in private as we are in public? How do we really feel about people whose culture is 'different'?

A couple enters a hotel elevator in Calgary and, finding it occupied by only one other person – who happens to be a friend of mine from Montreal – feel perfectly comfortable in starting a one-sided conversation that begins with suspicions of a "curry smell" in the hall and ends with the observation that most cabbies in Calgary seem to be "Pakis."

Later, the Montrealer finds himself in a cab whose driver is clearly not a "Paki" and who, after quickly establishing that his fare is an anglo, proceeds to rant against the "French" in Quebec.

These are not cabbie stories but examples of awkward

situations I suspect many of us find ourselves in as cultural diversity brings us into increasing conflict with our ingrained prejudices.

These are situations that are rarely of our own making – my cabbie, after clearing his doubts about my ethnicity, had no trouble dumping on Jews on the presumption that my being non-Jewish had created a bond. My friend's encounters also assumed ethnic and linguistic bonds that gave strangers the freedom to make nasty comments about others.

My friend and I both found ways of firmly disabusing our offenders of their misplaced kinship, but the situations and others that I have heard of beg these questions.

What do we really think of each other? And are we as noble and sensitive in private as we are in public? What are our true attitudes and perceptions about people whose culture, language, religion and ethnicity might be different from our own?

A few weeks ago, the National Conference of Christians and Jews made public the results of an exhaustive survey exploring these questions in the U.S., where multicultural diversity is beginning to rival Canada's.

Although racial and ethnic issues in the U.S. have a different historical context from ours, the results of that survey, with all their contradictions, are thought-provoking.

The main findings of the survey were somewhat predictable. Most blacks (80 per cent), Latinos (60 per cent) and Asians (57 per cent) agreed with the statement that whites are "bigoted, bossy and unwilling to share power." Majorities of the non-white groups also blamed whites for depriving them of opportunities in education, jobs and housing. In contrast, a majority of whites (63 per cent) felt that minority groups are given equal opportunities in these three areas.

There were signs of positive views in the survey. More

than 80 per cent of those surveyed said they admired Asian Americans for "placing a high value on intellectual and professional achievement" and "having strong family ties."

A majority of all groups agreed that Hispanic Americans "take deep pride in their culture and work hard to achieve a better life." And big majorities said blacks "have made a valuable contribution to American society" and "will work hard when given a chance."

Survey results that weren't so predictable were the negative attitudes minorities revealed about each other.

For example, 46 per cent of Hispanics and 42 per cent of blacks agreed with the statement that Asians were "unscrupulous, crafty and devious in business," 68 per cent of Asians and 49 per cent of blacks said Hispanics "tend to have bigger families than they are able to support." Further, 31 per cent of Asians and 26 per cent of Hispanics agreed with the statement that blacks "want to live on welfare."

Responding to statements about Jews, 54 per cent of blacks, 45 per cent of Latinos, 35 per cent of Asians and 27 per cent of non-Jewish whites agreed that "when it comes to choosing between people and money, Jews will choose money."

All groups endorsed affirmative-action programs and accepted the importance of cultural diversity in education and society. However, a set of questions produced these results:

Whites felt most in common with blacks, but least in common with Asians. Blacks felt least in common with whites and Asians and most in common with Latinos. Latinos felt least in common with blacks and most in common with whites. And Asians felt most in common with whites, who felt least in common with them.

While pondering the ironies and contradictions of this survey and the disturbing correlation between prejudice and burgeoning diversity that it appears to indicate, I found myself recalling a dinner years ago at a friend's home in Toronto.

A desultory conversation with my friends – a couple from India – about the difficulties of finding a live-in caregiver for their children took an unexpected turn.

"It's impossible to find a good nanny here," said my host.

"We're planning to bring one over from India."

"But the Toronto *Star* is full of nannies offering their services," I pointed out.

"Yes, but most of them are black."

"So what?" I asked, dreading the reply.

"Well, you know, they're lazy and they steal."

To this day, I feel I should have got up and left. Instead I chose to stay and protest, which merely made my friends change the subject.

I remember going home brooding about my inadequate response and the need for minorities to tackle their own prejudices while battling those they face from others.

More than 15 years after that dinner, I find I'm still in despair about its implications – a despair that can only intensify with the survey south of the border.

There's always hope

On certain nights, when the moon is getting fuller or there's been a particularly nasty spate of killings in the news, you go to bed weary of the world's woes.

Why do Serbs rain shells on hospitals? Why do Muslim and Jewish fanatics slaughter each other in the Middle East? What about Rwanda and Belfast and Kashmir?

> When religious practices are liberal and open, they can become a window on others, a pipeline of understanding and a buffer against the world's ills.

It seems that everywhere you turn, people are killing each other in the name of God and country and ethnicity. The despair generated by this hate-driven violence is becoming so endemic that we often don't even think about it, let alone talk about it.

Yet life seems to go on undisturbed, even for those who live in or are close to the theatres of violence. People still get married, have children, throw parties and go for picnics. Call it resilience or call it delusion – the show does go on.

And for those of us who live far from the violence or watch or read about it in the peace and comfort of our armchairs, life can take on a rhythm so smooth and seductive that you simply tune out the horror. Until the next big atrocity hits the headlines.

It was with these sombre thoughts last Saturday that I arrived at the Temple Emanu-El-Beth Sholom for the bar mitzvah of 13-year-old Jacob Harris, son of a colleague.

Temples, either Jewish or Hindu, and other edifices of formal religious worship are not places that I frequent. The farther I stay from organized and ritualized religion, the more comfortable I feel. This is not to say I detest religion; rather I abhor fundamentalism in religion, no matter which god it represents.

The exception, of course, is where religious practice is liberal and open, more a way of life and culture than a creed of exclusion and oppression. Then religion becomes a window on others, a chance to learn, a pipeline of understanding.

Jake's bar mitzvah offered a unique opportunity to field-test views whose popularity seem to be steadily eroding in the face of polarization and hardening of attitudes, religious and political.

Both Jake and the temple proved to be perfect symbols of the liberal tradition of Judaism.

Jake's mother is of Japanese heritage with a Buddhist and Christian background and his father is of European Jewish heritage.

And the reform temple, built in 1982, is open to all and women enjoy equal status. Indeed, in Tracy Shuster, the temple has a female cantor, a practice unheard of in orthodox synagogues.

Whether it was the spring sunlight streaming into the temple from the glass in its lofty, domed ceiling or the casual, welcoming ambience or Shuster's lyrical voice, I felt an instant lightening of my dark thoughts.

As Jewish traditions go, the bar mitzvah is relatively new.

In a religion that goes back at least 4,000 years, the coming-of-age ritual for boys (these days they also have one for girls called the bat mitzvah) dates back only to the 13th century.

Jake, who was entering the age of responsibility on this day, had spent a year studying and learning passages from the sacred Torah, the parchment scroll that contains the Five Books of Moses.

And, in studying for his new role as a young adult Jew who can assume religious responsibilities and observe commandments, he had come prepared to chant in Hebrew and offer a short resumé of how he intends to govern his future life.

Belying his years, Jake spoke and sang with a confidence that impressed all the invited family members and friends. And Rabbi Leigh Lerner, speaking colloquially and in English, managed to condense an ancient and complex doctrine into a user-friendly set of skills that can help people negotiate through life.

Using such contemporary metaphors as hockey and Hollywood movies, the rabbi put not just Jake, but everyone else, at ease as he led and sang prayers in Hebrew and preached in English.

For his part, Jake, sporting a crested, navy-blue blazer and grey flannels, cut a heroic little figure as he talked of loving and respecting one's neighbors with guidance from God, to whom he referred to as he or she and as someone who loves all man and womankind.

Moved by Shuster's inspired singing, browsing through translations of Hebrew prayers and the Torah and listening to shiny-cheeked Jake's brave vision of a loving and peaceful world, all lingering dark thoughts yielded to the infectious hope only the young seem to possess.

The thoughts and despair would doubtless return, but right then it was time for a celebratory lunch and a fervent toast to Jake's and all our children's future.

Africa's miracle

We are saying let us forget the past. Let us hold hands, as we have done here.

– **Nelson Mandela at a reconciliation service in Soweto yesterday**

We are the rainbow people of God. We are free, all of us, black and white together. We are free!

– **Archbishop Desmond Tutu at the same rally**

It hardly seems real that Nelson Mandela, this titan among us, is free, has survived all those years in prison and has steered his party and country to a sweet turn.

They had names like Maharaj and Suneeta and Sonny. They were among hundreds of South Africans who came to study medicine and engineering at Indian universities in the 1960s and 1970s.

The ones that came to the university I studied at – Nagpur – were all of Indian ancestry, but that's about all they had in common with us locals.

The South African Indians dressed differently (rather

more fashionably) and talked with different accents.

As with their clothes, their taste in music was far more Westernized, thereby creating an instant rapport with classmates in the English-language colleges they attended. This was the era of international student ferment and solidarity – whether on politics or the Beatles.

The South Africans were also a cheerful bunch, with sunny smiles and an infectious sense of humor, albeit dark, operation-room humor, a result of their ostracization at home.

There was a lot of pain and suffering behind those smiles. Fiercely patriotic about their homeland, they were also the innocent victims of the evil system of apartheid that segregated them from their countrymen and forced them to travel far afield for specialized education.

As college students who had been born in a free India, oppression and colonial persecution did not quite carry the same direct meaning for us as it had for our parents and grandparents.

But since the previous generations were still around to talk and write about it, we had all grown up in the heroic shadows (some of them posthumous) of people like Mahatma Gandhi, poet Sarojini Naidu, Jawaharlal Nehru, the Mahatma's political heir and first prime minister of India, and Babasaheb Ambedkar, the brilliant lawyer and leader of the Untouchables who drafted India's constitution.

Led by Gandhi, who had honed his ideas of nonviolent revolution in South Africa before giving full rein to them in India, these and other prominent freedom fighters had infused their peers and subsequent generations with the inspirational ideal of equality and freedom for all.

No revolution is a perfect one, and the saint-like Gandhi, a bibliography on whom is second only to that on Jesus Christ, had his detractors in his lifetime and has them today.

But for millions in India and around the world, the simple, loincloth-clad Gandhi was the epitome of principled and bloodless struggle against slavery and discrimination.

No wonder, then, that when Gandhi was martyred in 1948 by a fundamentalist-Hindu assassin's bullet even as he was in the midst of praying for ethnic and religious unity, another great man of this century, Albert Einstein, was moved to say: "Generations to come, it may be, will scarce believe that such a one as this ever in flesh and blood walked upon this Earth."

No wonder, too, that during his lifetime and the time he spent in South Africa and in its jails, a close bond had developed between the freedom movement there and in India, between the Indian National Congress and the African National Congress.

Indeed, the South Africans studying in India were a symbol of these special ties, daily reminders of the oppression our generation had mercifully grown up without, but that was still practised in their homeland.

But imbued as we were with the need for freedom and equality in South Africa, it was impossible not to despair about the plight of the disenfranchised there. If India's recent history had raised hopes of freedom for colonized people elsewhere, it had also proved how tenaciously, deviously and brutally oppressors can cling to their power when it comes under threat.

The clichés and contradictions raised and sparked by the regime in Pretoria over the years all held painful personal resonances for anyone who lived or had grown up in India.

And it hardly made a difference to listen and observe from the other side of the world, having moved to Canada in the early 1970s. Sure, there was a special love and interest here for South Africa and the struggle for emancipation there. And yes, like thousands of Canadians, I refused to buy or knowingly consume South African products.

But all along, the message that came through was

depressingly familiar. Blacks and other minorities were disunited and incapable of governing themselves. The country would collapse in ruin and bloodshed if apartheid were dismantled. And so on. For sheer paternalism and arrogance, it was hard to beat until you looked up what the British said and did in India.

But history – some would even say Nature or God – has a nasty habit of foiling the best-laid plans of evil men, coughing up from the depths of despair a single superhuman figure to change things around.

For millions in my lifetime and that of my parents and grandparents, Gandhi was such a figure. And so was Lenin and, until the excesses of the Cultural Revolution, so was Mao.

And so is Nelson Mandela, this titan among us, who, like Gandhi, has wrought what can only be called a modern-day miracle.

It has been a tortuous road and even now, with the new, inclusive South African parliament proclaiming him president today, and as the world's leaders arrive for his installation tomorrow, it hardly seems real at all.

It hardly seems real that Mandela is free, has survived all those years in prison and has steered his party and country to this sweet turn.

It hardly seems real that in achieving the impossible, he has opened his heart and arms in reconciliation to his detractors and former oppressors.

But it is all real, and I, for one, am overcome with the same tears that I have shed over South Africa through the years, right back to the days when some of its segregated citizens were my college-mates.

Only this time, they are tears of joy.

Fierce expectations

> Society often imposes rigid expectations on immigrants. But newcomers, like their hosts in the mainstream, are entitled to their fair share of dummies, crooks and criminals.

'OK, sahib," the porter would say, "I am now going to push you into the train. Don't worry about falling back or your suitcase. I'll prop you up and pass your suitcase through later."

And so this teenager, returning to college after summer holidays at home, would dangle in mid-air, tennis racquet in one hand, a tote bag in the other, as he was propelled, feet first, through the open window of a crowded coach.

There would be immense resistance from those inside, of course, but most of it would be verbal. Whatever physical opposition did take place would always be too feeble for the porter, a veteran of "placing" passengers in unreserved, open-seating coaches on Indian trains.

It was necessary to enter through the windows because it was a little easier to ram through them than through the doors, which were solidly barricaded with luggage and people.

The porter also knew that no one would seriously impede progress through the window for fear of hurting his client, particularly if the client was a nerdy, helpless teen with glasses.

The process would take all of the three minutes the train halted at this station in southern India with money changing hands only at the end of it all, often with the poor porter running alongside the departing train.

Inside, within minutes, a minor miracle would occur. Somehow, a small sitting space would be found for the intruder, and within minutes he would be engaged in animated conversation. Where was he from? Where was he going? What did he study? Would he like to share some food?

And, an hour or so later, such a strong bond would have been established that the newcomer would join his new buddies in trying to keep other window entrants out at the next station.

This train from the past keeps steaming through my mind these days, abiding metaphor that it is for the ebbs and flows of immigration. Sometimes, its chugging becomes painful as I catch myself muttering darkly over the latest horror story of crime or financial misdeeds by an immigrant, be it a professor or delinquent.

As a first-generation immigrant, I can't help feeling ashamed over such incidents, even though logic and facts dictate otherwise. We all know that all immigrants cannot be blamed for the crime or misdeeds of a few, that the overwhelming majority of us are as law-abiding as anyone else.

Yet, I find my embarrassment translating into a disturbing desire to put bars on those train windows or slam them shut and keep all newcomers out.

Such resentment is disturbing because it is so clearly self-serving. Having come aboard a crowded train and found a comfortable niche, it appears that I don't want others to share or threaten it.

Is this, then, how the journey should end? A final arrival that can only thrive on exclusion?

Surely not. There has to be another explanation. Isn't it a distorted sense of tribe and pride – the notion that someone's color, ethnicity, language, community or immigrant status should take collective blame for individual actions?

But migration, with its attendant baggage of rejection and adventure, struggle and success, frustration and celebration, seems to create such a fierce desire for acceptance in all newcomers.

It's a desire so compelling that it becomes a pervasive culture that just can't seem to cope with failure, with the misfits, with those gone astray. In this mode, it draws nourishment from the host society that often imposes higher standards and expectations of performance and behavior on immigrants.

But surely immigrants are entitled to their share of dummies and dunces, crooks and criminals?

What's more important, the crime, its causes and prevention, or the origin of the perpetrator? After all, today's immigrants are tomorrow's hosts, aren't they?

So we need better screening of immigrants and enforcement of laws – and who can argue with that?

But how can you possibly predict that the teenager being pushed through the train window is not going to fit in the crowded coach? Or know that an 8-year-old boy entering the country is going to grow up to be cop killer? Or guess that emigré destined to be a professor is going to end up murdering four of his colleagues?

Somewhere you have to use trust and judgment.

Somewhere you have to factor in the human condition as it affects us all.

Somewhere you have to step off the oppressive treadmill of expectation and achievement and let the journey unfold on its own.

A turn with Mama

As the lights dimmed, the DJ put on an Italian lament and invited the bridegroom and his mother to take the floor.

The words were a mystery to me, but their intent seemed obvious. A young man had come of age, was getting married and leaving home. A mother, happy and heartbroken at the same time, was "losing" a son.

> There is something timeless about a mother's love, an emotion that transcends all cultures and boundaries. But sometimes we realize it only in intense moments.

At least that's how the song affected me as I watched my friend Riccardo Spensieri, clutched tight in his mother's arms, shuffle across a floor cleared of everyone else for this dance.

The guests carried on with their merrymaking. One of Rick's buddies let out yet another joyful, but annoying, toot on a compressed-air horn. Little girls and boys carried on running and chattering all over the hall. And the bride, Heidi Larsen, watched with a gentle, happy smile from the head table.

But were those tears in Mrs. Spensieri's eyes as she gave

her son a sudden, fierce squeeze?

And were Rick's eyes moist as he responded, somewhat awkwardly, with a kiss to her forehead?

It was hard to tell in the soft lights. And I wasn't sure of the filter I was seeing all this through.

There is something timeless about a mother's love, I thought. An emotion that transcends all cultures and all boundaries.

Yet the realization of this essential and universal facet of our lives only seems to come in moments of intensity – oh, Mama! we cry in times of pain, but also at moments of joy.

And outside these extremes, we seem to devote so much time to denial and often to downright abuse or rejection.

"Only a mother can understand what goes on in a mother's heart," my mother used to say, almost to the day she died.

To which, whether as a troublesome teen, established adult or returning prodigal, I would generally respond with some scorn, misreading the sentiment as maudlin or perhaps manipulative.

What made it so hard to accept a mother's affection at face value?

Why do so many of us, especially men, get all awkward at any public display of such affection?

It would be easier to combat if the interaction, at least the negative aspects of it, stayed within the confines of a particular relationship or circumstance.

Unfortunately, in some cumulative, subconscious way such interactions seem to spill into the public domain, affecting our attitudes about not just one's mother, but any woman who has a child and, inevitably, all women.

Again, such attitudes are not restrained by geography or culture, although their impact might be tempered by progressive legislation and enlightenment in some countries.

But whether it's Canada or India, Japan or Zambia,

Russia or Brazil, women continue to be denigrated and exploited, socially and economically. And within this framework, mothers and women contemplating motherhood face unrelenting hostility and pressures in areas like birth control, child care, career tracks and job security, to mention only four.

Yet, through it all, in repressed or liberated mold, women carry on, bearing their sons and daughters, ignoring their tantrums and follies, crying and laughing with them, helping them – as much as they're allowed to – through life's eternal tangles.

Ridicule, rape, murder, contempt, torture – women and mothers have experienced it all in measures far beyond most men's experience and imagination.

"How can I explain to you in mere words the mamta that fills all mothers' hearts?" my mother also used to say.

Mamta, an all-encompassing Hindi word derived from Sanskrit, tries to express in two syllables all the love, compassion, adoration, care and forgiveness that mothers embrace their children with.

Children who never really understand until it's too late or they happen to glimpse a flash of it during an idle moment at a wedding.

ASHOK CHANDWANI

Buntys and Pinkies

When I first heard about Terry and Sherry, I assumed they were somebody's pet cats.

It came as a distinct shock to learn that they were twin brothers whose real names were Trilochan and Surinder.

Like many upscale, westernized Punjabis in India and abroad, the twins had been given easy, if somewhat quaint, names. And, in this case, gender-neutral ones.

> When people move to a new country, sometimes the baggage they bring includes tongue-twisting polysyllabic names. The results can be unpredictable.

I had been boarding with a Sikh family while at university in the '60s when I first met the dashing, turbanned twins who tooted around the city of Nagpur in a Triumph convertible.

The family I lived with had its own way with names – one of their daughters was called Happy and a niece who was my professor of English was known as Cookie, although it was written as Kuki.

Through these Punjabis, I met a bewildering range of Sweeties, Buntys and Pinkies, all of whom added a fresh dimension to the mind-boggling variety of names in India.

As the child of refugees who were far more mobile than the locals, I had gone to school with kids with such easy first names as Raj, Meena, Ajay, Bindu and Nilu and last names such as Roy, Dubey, Das and Singh.

But I also had been exposed to such polysyllabic twisters as Thyagrajam Venkatacharya Pillai, Veernarayan Mukutophadhaya and Sherniwaz Jhoonjhoonwala. Inevitably, and with the consent of the owners, such names became Venky and Muku and Sheri.

Mercifully, my own name, Ashok, as common in India as John in Canada, had escaped any alteration, except for unsuccessful attempts by an affectionate aunt to call me Ashu.

And in polyglot India, everybody seemed comfortable with Chandwani except my Nagpur landlord, who to this day calls me Chandu Halwai after a famous confectioner in Delhi.

All this has changed since the day I arrived in Canada, where my name has taken on a significance it was never meant to have.

Put simply, for mainstream English and French Canada, my name defines and reinforces my "foreignness"; it's a dubious label that thwarts any attempt I might make to assimilate.

Now many an immigrant before me has shed this ultimate baggage and "adapted" his name to make it familiar and palatable to the mainstream. But for every Irish person who has dropped the O from his name or every Hariprasad from India who has become Harry, there are many like me who refuse to do so.

What you believe, eat and wear can change. How you walk, talk, dance, greet, kiss, toast or mourn can also change. But your name? Isn't that sacred and inviolable? The only link to your past, your family and your ancestors?

The only certain way to define who you are, over and above the flimsy mantles of ethnicity, language and nationality?

You have only to look at the nation's phone books to see how diverse Canada has become. Sure, you'll still find hundreds of Smiths and Tremblays. But you'll also find a staggering variety of other names, including such megasyllabic ones as Vandemeulbroocke, Vandenieuwegiessen,Hadjiathanassiou, Vengadesasarma, Zabiholahizadeh,Thamotharampillai, Theodorakopoulos, Khachkhechvan, Xoumphonphakdy, Zaccagninomoroni, Papadimitropoulos and Demechtchiouk. And these are only the family names!

I dare say that many with names like these have cheerfully accepted condensed versions or nicknames from friends. But I would also suggest that, like me, they are fiercely attached to their real names, or else they wouldn't print them that way in phone books or drive credit-card companies or licence bureaus crazy with such insistence.

And the least they – we – would expect from strangers is an honest effort to study and untangle the syllables. Hell, people could at least ask how to pronounce a name before proceeding to mangle it!

If they did, they'd discover that these weirdos with unfamiliar names aren't nearly as uptight as some of those seeking to pronounce them.

Pride aside, I'm constantly amused by how both my names still stymie almost everyone I know or meet. And the names clearly baffle strangers who might be phoning or sending me the mountain of junk mail that is every journalist's burden.

Over the years, people have been as stubborn in their reluctance to apply the simple rules of phonetics to my name as I have been to anglicize it.

Curiously, the problem seems unique to English speakers because francophones and allophones, by and large, haven't had any notable problems except for one mailing that consistently calls me Chandwanionher.

As for anglos, well, they've been known to refer to me as Eh-Shock, Ah-Shock, Ass-Hock, A-Shock, Ash-uk, Ashook and even Shock.

Others, out of mutual affection and consent, call me Ashnook, Schnook, Schookie, Schnookers, Shokie, Ashy and Ash, the last also being a favorite with bartenders and politicians whose professions force them to be cordial and correct the first time they meet you.

Living in Quebec has also added its special flavor to this game (for that's what it has become for me now) and even led to a complete change of name once. This was years ago, when former Liberal leader John Turner was being introduced to a *Gazette* editorial panel. Having breezed through a Mark, a Henry and a Jim, his eyes betrayed only a flicker of confusion at Ashok – "Hi, Jacques," he said, pumping my handing heartily before moving down the table.

This was the first and only time that someone has francicized my name – an experience far preferable to caninizing it. This, too, happened years ago, when Bill Turpin, a buddy and colleague with an incorruptible first name, fielded a phone call on the *Gazette* news desk.

"Yes, ma'm, he's right here," said Bill, a mischievous gleam lighting up his eyes.

"I'll hand the phone to him right away, ma'm."

And, covering the mouthpiece so the caller wouldn't hear him chuckling, he said:

"I think this is for you, Mr. Chiwawa Dandy."

Serves me right, I suppose, for having thought of Terry and Sherry as cats.

A lousy year

A bearded priest wearing a crown-like embroidered hat with a photo of Jesus arrested my channel-surfing fingers over the weekend.

The television broadcast, in Arabic on an Egyptian community program, could just as easily have been in Greek, Russian or Ukrainian – Jan. 7 was Christmas for many Orthodox Christians who follow the Julian calendar.

> Sure, some got promotions in 1994. Others became proud parents. And love blossomed for many. But over-all, there was plenty of evidence that many human beings parted with their senses.

I couldn't understand the words, but their purport seemed clear – peace and prosperity in a caring world. Two weeks ago the same message had been broadcast by other priests and even politicians around the world.

And last week, millions of revellers from Tokyo to Manhattan, including a few score at my home, had ushered in a new year with champagne, chocolate, fireworks and words of hope.

But during the parties, and now, watching the declamations from the serene face on the tube, I found it hard to ignore the underlying hollowness in all the exhortations and exultations.

Let's face it, 1994 was a lousy year. Sure, some of us got promotions. Many became proud parents. And love blossomed for others. But in the larger scheme of things, human beings revealed more evidence of having parted with their senses.

In Rwanda, a fratricidal war claimed more than 300,000 lives. That's not quite the number of Londoners who danced in the fountain in Trafalgar Square, but it's certainly about the same number of people who crowded into Times Square in Manhattan to watch a huge aluminum ball studded with 180 white lights slide down a flagpole.

The Rwanda conflict, which also spilled into neighboring Burundi, was only one of about 70 conflicts that raged around the world last year, according to an Associated Press report. These included the high-profile wars in Bosnia and Chechnya and a little-known insurgency in Sierra Leone.

Only a few of these conflicts are political in nature – the rest can all be blamed on religious fundamentalism, ethnic nationalism or tribalism, often an explosive combination of two of these or all three.

The Russians have long left Afghanistan, but interclan and tribal rivalry continues unabated. The Russians are now entrenched in another deadly battle in Chechnya that has already claimed thousands of lives.

Religion hasn't become the rallying cry yet in Chechnya, but judging from certain noises there and in neighboring regions, it soon might.

There is no such ambiguity in Algeria, Egypt and India, where Muslim fundamentalists are key players in political battles that have claimed thousands of unnecessary lives.

Precise figures are hard to come by, but more than 17,000 civilians and military personnel have died in the past five years in Kashmir and more than 11,000 in Algeria in the past three years. The toll is nowhere as high in Egypt, but the violence is just as deadly and vicious from all sides.

In all three countries, secular-minded governments are in conflict with Muslim fundamentalists, except in India, where Hindu fundamentalists are also a factor.

But it would be facile and stupid to blame it all on Islam. After all, millions of peace-loving Muslims around the world feel equally distressed about such violence, as do millions of peace-loving Hindus, Buddhists and Sikhs about killings and atrocities in India and Sri Lanka.

And you only have to look at Serb atrocities in the former Yugoslavia to add Christians to this list.

Sarajevo, long considered a model for secular coexistence in Europe, is ruined, its multicultural splendor permanently stained by vicious ethnic cleansing.

No, the unavoidable truth is that wherever you turned in 1994, people seemed bent on killing each other for some xenophobic or zealous cause, striking severe blows against tolerance and cosmopolitanism.

And, I thought gloomily, flicking from the inspirational images on the Egyptian channel to news clips of burning buildings in Grozny, it's not going to be any better in 1995.

Home-cooked joy

"So, you're, er, adjusted?" asked a smiling George Chan, who runs a variety store-cum-deli with his wife, Jenny, in LaSalle.

The unexpectedness of the guileless question made me hesitate slightly before I nodded a vigorous Yes. And it sparked a catalytic reaction that stretched across two decades and Canada.

> Early exposure to the varied cultures and cuisines in Canada has saved me from falling into some cultural or physical ghetto, a fate that befalls far too many newcomers.

A friend and I had navigated through thick fog and the somewhat disorienting geography of the Montreal suburb to the Chans' store in search of home-cooked pierogis.

"Tell them Luba Semenak's daughter referred you," a colleague had said.

The words, repeated to the Chans, had acted like a magic incantation, opening a sunny window of warmth on a day that was as damp as the fog enveloping it.

Basking in the glow of this referred encounter, we had discussed the best way to defrost the frozen pierogis, the merits of their smoked bacon ("here, taste a slice first") and where to shop for authentic smoked pork loin.

And tickled by the notion that I was hosting a belated Christmas dinner party for a friend of Ukrainian heritage, George Chan, also of the same heritage, had wondered about mine and the year of arrival in Canada.

Having arrived here himself in 1947 – "after the war, when I was in a labor camp" – Chan seemed a bit surprised that I had only been here since 1973, leading him to ask the question about adjustment.

Driving home and through most of dinner, the question stayed in my mind, uncorking memories of those first weeks in Canada in Vancouver, including one – I'm not making this up – of pierogis!

Within days of arrival I had found a job with the Company of Young Canadians, charged, with two others, to try to start a community newspaper in an immigrant neighborhood.

So, there we were, these three fledgling Canadians, a man from Italy, a woman from Hong Kong and me, from India, struggling to raise money for a newspaper whose main goal would be to help immigrants adjust to Canada.

In hindsight, the exercise was as doomed to fail as the federally funded CYC was to disband. But at 23, fresh off the boat, it was my first enduring taste of multiculturalism – in thought and in deed.

During the six short weeks I spent in Vancouver, not only did I acquire a lasting love for things Italian and Chinese, but also Ukrainian, the last because of frequent lunch visits to a tiny restaurant that specialized in pierogis.

If an adult immigrant's arrival in the new country is the beginning of a new life, then surely the influences of the first few weeks or months are as formative as those at birth in the old country.

This was certainly the case for me – that early exposure to varied cultures and cuisines saved me from falling into some cultural or physical ghetto, a fate that befalls far too many newcomers.

And, over the years, this early grounding has opened doors into a cosmopolitan world that would be unimaginable in the real ghetto – the Old Country.

The paths to these doors haven't always been straight and smooth, but they have been clearly marked and inviting.

Today, the people I have met and befriended behind these doors have roots in more than a score of countries, but their heads and hearts are firmly cemented in Canada.

Gratitude can make one sound sucky, but in a very real way, I find myself agreeing with Prime Minister Jean Chrétien when he says that we live in the best country in the world.

Gobbling pierogis and gazing at the flickering candles and wine glasses, a symbolic sheaf of wheat and nine happy friends at my Ukrainian celebration, I couldn't help but nod again to George Chan.

If happiness is the test, I thought, then I guess I really have adjusted.

Airborne's legacy

Speaking about Canada's role as international peacekeeper, Gen. John de Chastelain, the nation's top military officer, told a U.S. audience last week that this postwar tradition stems from our national personality, which he described as multicultural, built on tolerance, "even bland, even decaffeinated."

> The video saga of racist and sadistic behavior by our 'elite' soldiers raises deeper societal questions. Are we deluding ourselves about our national personality traits?

The short wire-service report from Bowling Green, Ohio, where our chief of defence staff was the featured speaker at a university symposium, did not elaborate on this short quote from the general.

But for anyone following the video saga of sadistic and racist behavior by some members of the Airborne regiment, the general's remarks underscore a key issue in the national handwringing about the misdeeds of these "elite" soldiers.

This issue is that of our self-image, and, by extension, the level of self-delusion required to sustain this image.

A measure of this problem is the range and number of

apologists defending the behavior of the errant soldiers. The distorted logic offered by some of these pretenders is as sickening as the acts of the soldiers.

That this indulgent "boys will be boys" behavior stains the broader, impeccable honor and traditions of Canada's military and might violate the new military regulations and laws seems lost on this group.

And to suggest that consent can condone or mitigate the seriousness of offensive public acts is ludicrous, because it undermines the validity of any social contract based on civility and respect for the dignity of individuals.

Clearly, what people do or say to each other in private is their business. But that's hardly the context and issue in the Somalia and hazing videos.

No, the actions of the soldiers on these videos loudly and painfully suggest that there is a certain smugness about our assumptions about ourselves.

Having adopted some of the world's most advanced legislation on human rights, we sit back and expect a magic wand to remove or alter the conditions that necessitated it.

Having decided that, after their valiant role in the world wars, our soldiers will be peacekeepers to the world, we expect them to instantly become models of culturally sensitive diplomats.

If only it were that simple to change old habits and behaviors! It would save us from reacting with such self-righteous polarity each time some ugly reality revealed itself.

This is not a defence of the Airborne soldiers, but disturbing as their actions are, don't you find it alarming that many among their officers and the public are surprised by them? I mean, really.

The point here is that tough as it is to inculcate notions of virtue, valor and decency in soldiers and the public alike, it is far more difficult to impose concepts of peace and tolerance.

Any discussion, then, of the multicultural, tolerant Canadian society that de Chastelain alluded to has to ultimately examine closely the role of its military. This is not a naive peacenik's suggestion, but a reasoned take on the changes that have occurred since the Cold War ended.

Everything that's happening to Canada, from demographic changes to shifting trade alliances, supports a scenario of Swiss-like neutrality and a reduced military devoted to international peacekeeping and such "civilian" functions as rescue and disaster relief.

This might sound like disbanding, or to borrow from de Chastelain, "decaffeinating" the military. But it's not. It's more a question of focus and bringing the military into step with the direction Canada seems headed in and the impulses guiding it.

This wouldn't be the first time in history that a society changed its military orientation.

The idea of a pacifist, non-combatant world has been explored by many an enlightened emperor and philosopher – from India's Asoka in the third century B.C., who renounced war after a particularly bloody conquest, to war-weary Greek intellectuals in the fourth century B.C. And later by those who controlled empires such as the Roman, Spanish and British.

Success in establishing a durable peace has been sporadic and not all that long-lived, underlining the paradox that to make peace you have to make war.

This was something that de Chastelain also appeared to allude to in Saturday's speech, whose theme was Peacekeeping as an Expression of Canadian Values.

The Associated Press quoted de Chastelain as saying: "We (in the military) believe you train for war and if you train for war, you don't have to fight them.

"We do train for war and as a result, we can keep the peace."

The general will doubtless echo this theme again today as he presents his report to Defence Minister David Collenette on those disgusting Airborne videos.

But if the only result of that report is severe disciplinary measures against the implicated soldiers or even the disbanding of the entire regiment, then we will have addressed only the immediate problem.

The larger issue of reconciling our self-image to reality as a prelude to achieving that elusive ideal of a truly just, peaceful and peace-minded society is going to remain with us for quite a while.

A confusing battle

> Immigrants are predisposed to avoiding discussions about sovereignty and separation because they raise the spectre of the kind of economic and political 'troubles' they have left behind.

The battle for Quebecers' minds has begun.

The pawns are ready for sacrifice.

The puppet-masters are poised.

Pick the metaphor you like or write your own as Quebec's roving commissions begin public consultations today on the Parti Québécois's draft bill on sovereignty.

How about this one: the sovereignty cauldron is bubbling over, threatening to anoint us all in the sacred steam of self-determination.

Sounds absurd and cynical? It is.

While I'm not among the 24 per cent of allophones in a recent poll who found Premier Jacques Parizeau's sovereignty plan confusing, I can certainly understand why they and other respondents feel that way.

In one sense, it's inevitable that many allophones, particularly first-generation immigrants, find Parizeau and the PQ's pronouncements and manoeuvrings confusing. This is because, over the decades, very few immigrants have come to Canada to fight for or against Quebec sovereignty. It's hardly the kind of cause that gives birth to international brigades of revolutionaries. Or mercenaries, for that matter.

Sounds simplistic, but it's true and easily verifiable that whenever immigrants have come to this country, it has been either for economic reasons or because they're fleeing some form of political oppression.

What I'm saying is that immigrants are predisposed to avoiding discussions about sovereignty and separation because they raise the spectre of the kind of economic and political "troubles" they have left behind.

This doesn't mean that allophones are ignorant of the issues or that they have no opinions. No, they're merely reluctant, sometimes out of an almost courteous sense that this ain't their problem.

But let's agree that this is a shortsighted attitude. Let's also say that having lived here long enough, immigrants and their children and grandchildren have become assimilated enough to participate in such discussions. Let's even say that they are keen to take part.

The questions then are: Are the conditions ripe for such participation? Are the exhortations and pleas genuine?

That these are doubts that cross many cultural and linguistic boundaries is obvious from the seeming confusion among members of the coalition representing the Greek, Italian and Jewish groups in Quebec.

The coalition, whose member organizations collectively represent about 400,000 allophones, had initially announced that it would boycott the sovereignty commissions because it felt the process was not legitimate.

Now, to applause from Parizeau, the coalition has decided to take part and offer a federalist viewpoint, even though it still feels the consultations cannot be considered legitimate. It's difficult to understand how one can contribute effectively to a process one considers severely flawed. But it's easy to understand that the coalition had no choice but to reluctantly participate and be seen to gain political legitimacy for allophones' interests.

The coalition's flip-flop further reinforces the self-professed craftiness of Parizeau's plan. You're damned for being isolationist if you boycott and you're also damned for the same reason if you do participate and support federalism.

A further problem, of course, is the contradictions and gaps in the PQ plan and the leadup to the hearings.

Consider Parizeau's recent visit to France in search of support for Quebec sovereignty, which I see as cause for unease in all Quebecers, be they committed federalists or dedicated separatists.

If, as the PQ has been claiming lately, a sovereign Quebec will be an inclusive and pluralist country, then seeking France's support only implies the opposite – that this whole exercise has something to do with the language and common heritage of the majority. Hardly reassuring for minorities, is it?

It was also in France that Parizeau made a curious statement while reacting to the news that a group of French artists had expressed their support for Quebec sovereignty. Here's what the premier was quoted by the Canadian Press as saying:

"Yes, indeed, I gather we have friends. . . . While driving for the sovereignty of Quebec, sometimes in Canada it seems to provoke antagonistic reactions, and that's understandable. It is so remarkable in the old Europe, and in the new Africa and in the eternal Asia to find support and understanding."

Even discounting contextual and translation flaws, Parizeau seems to be equating his party's struggle with the

anti-colonial independence movem...
tions like Britain, France and Holland ha...
in the past 100 or so years.

I really have to stretch my imagination to t...
point to accept that Quebec is indeed suffering the
imperialistic domination, exploitation and oppression ...
spurred independence movements in African and Asian
countries.

But let's say I'm misreading the whole thing; that I am profoundly ignorant of Quebec's past; and that I'm naive and stupid to believe that the politics of grievance and revenge can be overcome by negotiation and sensible redress.

I would still be far more comfortable responding to nationalist aspirations in their home setting rather than played out abroad. I would still rather see Parizeau and other sovereignists raise their flags, voices – even fists – here at home than watch them fawn over support in France.

And I say this acutely aware, as no doubt members of the allophone coalition are, that the draft bill does not even mention allophones, let alone any guarantees for them. The only minority guarantees offered are limited ones to the English-speaking community and aboriginal peoples.

Time to look for a new metaphor, I guess.

Terrorism's toll

A s the body count and horror mounts in the Oklahoma bombing, so does disquiet at how easily a lot of us assumed that it was the work of Mideast (read Muslim) terrorists.

Granted, there is history here. In recent times, some fundamentalist Muslims have been

> The bombing in Oklahoma resulted in some distressing finger-pointing at Arabs, 'Middle Eastern types,' Iranians and Muslims.

responsible for or implicated in so many high-profile terrorist acts that it is inevitable that some of this notoriety rub off negatively on the millions of law-abiding and peaceful Muslims.

In such a climate, it has also not been easy for others who share an ethnic, linguistic or geographic heritage to escape being stereotyped by non-Muslim terrorist groups active around the world – groups who nominally represent all major religions and all manner of political beliefs.

A short visit, phone call or query to a database is all you need to discover what Bill Clinton's government had finally begun to acknowledge before the Oklahoma blast confirmed it so murderously – that terror is a broadly based weapon used by militants of all persuasions.

According to published analyses in the American press, the Clinton administration has been endeavoring to change its foreign-policy initiatives to treat militant fundamentalism as a separate issue in its dealings with predominantly Muslim nations.

The Oklahoma blast revealed some evidence of the reasoning behind this long-overdue approach in the many statements and briefings that followed the bombing – several senior officials were extremely cautious in their speculations about the identity of the murderers.

However, their caution seemed to have scant effect on people and the media, whose responses appeared to reflect gruesome glee in automatically pointing the finger at Arabs, "Middle-Eastern types," Iranians and Muslims.

When it became clear that the suspects didn't match this stereotyping, I'm certain there must have been some others like two of my white friends, who exclaimed, "Thank God they were white Americans!" or words to that effect.

But this response, heartening as it might sound, did little to mitigate the horror of what happened in Oklahoma and, in a much smaller way, the subsequent experience of individuals like Abraham Ahmad, the Jordanian American who was interrogated at London's Heathrow airport and returned home to the U.S. in handcuffs as a false suspect.

Ahmad, a U.S. citizen from Oklahoma City who had been on his way to his native Jordan when arrested, told the Associated Press yesterday that he thinks his Middle Eastern looks and name and the fact that he was coming from Oklahoma City prompted authorities to detain him.

"People automatically think that the person who did this is from the Middle East," he said. "But I didn't think that the FBI would think so."

On one level Ahmad's plight and that of others – a Canadian of Iranian descent claimed yesterday he was unceremoniously picked up and questioned for an hour in Toronto last Thursday – could be attributed to the rumor-mongering that inevitably follows a tragedy of this magnitude.

But it is also symptomatic of the persistent xenophobia about immigrants, foreigners et al, with all its attendant, nostalgia-ridden rhetoric of decrying the loss of mythical "pure" values and lifestyle.

It comes as no surprise that, to date, the prime suspects in the unforgivable killings of innocent babies and office workers appear to have links to ultra-right groups obsessed by irrational, fanatic agendas.

What is chilling is the scale of the attack and the ease with which it appears to have been carried out.

If it turns out that the killings were part of a wider conspiracy, then there are further frightening implications. The questions we all have to ask are these:

Is the growing conservative assault on governments on both sides of the border – and its increasing public acceptance – creating a climate in which fringe groups from the radical right feel emboldened, if not empowered?

The Newt Gingrichs and Pat Buchanans and groups like Human Life International would be vehement in denying this, but why is their demagoguery finding so many friendly ears? Is it because in its failure to arrest growing income gaps and economic polarization, and its reluctance to streamline welfare and subsidy policies, the liberal left can only respond with a paralytic, intellectual vacuity once considered the hallmark of conservatives?

The answers to these questions will probably trouble us long after the perpetrators of the Oklahoma bombing are prosecuted and punished. But as the discourse continues, we all must pause and consider that hatred is on the rise all over the world, unfettered by any particular religion, creed or country.

What are we going to do about it?

Perils of denial

A front-page story in last week's London Sunday Times makes dangerous mockery of the sombre commemorations taking place around the world to mark the end of World War II in Europe.

Under the arresting headline "WINSTON WHO? PUPILS CANNOT NAME WAR HEROES," the story, quoting a survey of 1,600 state school students, exposes startling gaps in British schoolchildren's knowledge of wartime figures and events.

> Surely the greatest crime is to deny or justify holocausts. It's a crime that we can all legitimately be accused of abetting by ignoring Holocaust deniers.

According to the report, 36 per cent of children aged 11 to 14 failed to identify Winston Churchill. Another 25 per cent didn't know who Hitler was. And more than 60 per cent did not know what the Holocaust was.

Coming on the heels of a story the previous week in *The Gazette* that revealed that students who don't take optional courses in world history can graduate from high school in Quebec and Ontario without ever learning anything about the Holocaust, the London report undermines the fundamental significance of the weekend's anniversary.

If there is one feature that distinguishes the war whose end 50 years ago is being marked in ceremonies across the world, it is the naked evil it has come to symbolize for mankind.

That more than a mind-numbing 60 million soldiers and civilians died in the war is horrifying. But worse is the hate and evil that spurred this catastrophe, resulting in the ultimate shame of the Holocaust.

Millions have been butchered in other wars before and after this one. And millions of others have perished in ethnic and religious fratricide, victims of genocidal governments, groups or conflagrations.

But surely it is not an insult to the worldwide victims of other genocides and pogroms – before, during and since World War II – to say that the Jewish Holocaust is unique if only because of the cold-blooded and systematic manner in which it was planned, carried out and recorded.

Surely remembering or perpetuating its grim lessons in no way devalues the equally grim lessons learned from other hate-driven killings.

Surely the greatest crime is to deny or justify this and other holocausts – a crime that we can all legitimately be accused of abetting by ignoring Holocaust deniers or being apathetic.

The dominant message in the weekend's ceremonies was one of reconciliation among the nations involved in this horrible war. This is unquestionably an important and vital message for all of us and particularly for the nations who were officially clustered as the Allies and as the Axis powers.

These nations have been learning to coexist peacefully with increasing, though yet incomplete, success.

And many people born in these nations and their descendants have also been learning to live peacefully together in countries like Canada and the United States, where they have established new lives.

It would certainly be foolish for Canadians to interact with each other on the basis of which side their forebears were on during the war.

However, it would not be foolish at all to recognize that reconciliation and peace cannot be sustained by forgetting or being apathetic to history.

It is only by remembering and confronting the evil in our past that we can find and entrench ways to prevent or curb it in the future.

It is appalling that only 50 years after this cataclysmic war, there is so much ignorance and indifference about it.

As democrats, we are obliged to suffer the misinformed and misguided bleatings of those who deny the true lessons of the war and the Holocaust.

But we do not have to suffer in silence. We have to raise our own voices and counter these distortions every step of the way.

The best place to begin is with our children, be they at home or at school.

This is not a matter of indoctrination, but of healthy, informed discussion that can only increase the odds for a peaceful future.

Roots of paranoia

It's 1982 and I'm sitting with 20 others in a van designed for 12 and sweating, not from the blazing sun, but from fear.

My fellow passengers are all Muslims. Some of the women are wearing hijabs or are fully veiled.

> What is it in so many of us – in 1995 at that – that makes us shriek in alarm at the sight of a hijab or other religious headgear?

Everyone is speaking a language I know, but I'm so scared that I don't want to speak.

Everyone is joking, but I'm so tongue-tied that I can't laugh – to do so would reveal my secret.

Yes, I'm here in a self-imposed disguise, at the airstrip of a dusty little place called Mohen-jo-Daro, not far from the excavated 4,000-year-old site of the Indus Valley Civilization in the province of Sind in Pakistan.

This is the region that my parents had fled in 1947 when the Indian subcontinent was partitioned into a new Islamic state called Pakistan and an independent secular republic called India.

Being born after these events in India saved me the immediate violence, pain and suffering of forced dislocation. But it did not spare me any of the anger or nostalgia harbored by my parents and the 2 million or so other Hindu Sindhis who became part of a new worldwide diaspora.

It only seems fitting then, that in 1982, armed and transformed by a Canadian passport, I should find myself on a shaky roots mission, surrounded by the people whose religion my parents had always seen as the cause of their misfortune.

The problem is these Muslims don't at all look like the monsters my folks in India had painted them to be.

In fact, the more I look at them and the more I listen to them as we wait for the van to depart the tiny airport building for the larger town of Larkana, the more I feel embarrassed that I've chosen to keep my Sindhi identity secret for fear of trouble.

But since I cannot suddenly burst into Sindhi after pretending I know only English and Urdu, I'm forced to sit there and squirm at my unfounded fears. For as I listen to the chatter around me, the only danger I perceive is from myself – I'm terrified I might lose control and burst out laughing at the infectious humor emanating around me.

So much for demonization, I think, as I listen to banter about the government and sardonic speculation as to when the last passenger, an airport official who had come in the van to receive our flight, would show up for the van's return trip to town.

The jokes about officialdom and bureaucracy are nuanced by the fact that the official happens to be a woman, challenging, then and today, another stereotype. If they veil and oppress women in Muslim countries, how can they make them airport officials?

Interaction on a crowded minibus in one Muslim nation can hardly become the basis for an epiphany on prejudice. Neither can it become a defence or apology for the social and political ills affecting other levels and spheres of the country.

But, having been to many other Muslim nations since, and met and befriended many Muslims there and in secular nations, I do see this little minibus misadventure of mine as a personal metaphor for how easy it is to label, stereotype and ultimately demonize Muslims.

We all know that terrorism and fundamentalism are always the handiwork of a few and not confined to Muslims – every religion on Earth has its share of these evils.

Yet I am constantly amazed at the depth of the paranoia and prejudice that surround Islamic issues in Canada, a country whose constitution, legislation and top courts solidly commit it to the respect and accommodation of minorities.

What is it in so many of us – in 1995 at that – that makes us shriek in alarm at the sight of a hijab or other religious headgear?

What is it, if not irrational fear and suspicion, that forces groups like the Centrale de l'Enseignement du Québec to call for a ban on the wearing of hijabs, skullcaps, turbans and other religious headgear in public schools?

If the province's largest union of teachers – yes, teachers! – feels this way, then what hope is there for the rest of us to temper our insecurities with tolerance? To understand that there are far bigger threats to Quebec's and Canada's well-being than the headgear some believers might choose to wear?

One day, I dare say, some of us will find some enlightenment in a minibus somewhere else.

Until then, perhaps the least we could do is board a big bus right here at home and talk to some of these hijab- and turban-toting threats to our culture.

It just might change the language some of us speak.

Trail of rejection

> Do Canadians have the right to hurdle-free visits from relatives abroad? What happens to this right if the relatives happen to be from a country that's a source of illegals?

There are moments at Victor Sumbly's bustling Côte des Neiges pharmacy when you could be forgiven for thinking that he is one of those four-armed deities you see in Hindu temple art.

The Kashmir-born pharmacist's arms seem to be everywhere as he greets, embraces, kisses, shouts orders, answers queries and offers clients soft drinks and "the best prices in town."

"I'll be right there with you in a minute. Sit down and rest your feet for a while!"

"Here, take some Vitamin E home, it's an antioxidant and it'll help you with your cholesterol!"

"Bella, honey, what's my cost on Prozac?"

Looking at the blue-grey eyes shining from a face tanned from birth by the Himalayan sun, you could also be forgiven for momentarily thinking that Sumbly's on some happy drug himself.

But, no, the Prozac is for a couple looking for a deal on its price, something Sumbly has already cheerfully promised them.

"That's the least I can do for my family," he beams, including everyone in the store in his smile, and emphasizing ever so slightly the word family.

But for those who know him, this is hardly a surprise. Family's a word that keeps tumbling out of Sumbly as frequently as the words God, warmth, service and duty.

In the noblest philosophic traditions of his religion, Sumbly, a Hindu, has concocted a personal formula of ideal behavior and expectations that can be as infectious as it is intimidating.

It's a formula that is rooted in the universal precepts of sharing and community. And an unshakable faith in the sanctity of family – a large, loving and extended family.

It's also the formula that has sustained the 38-year-old Sumbly in the 17 years he's lived in Canada.

These are the years that have seen him earn his pharmacist's licence after graduating from the Université de Montréal and buying his own business.

These are also the years that have seen him marry Shakti, a research biochemist, and move into a comfortable bungalow in a tony, leafy neighborhood.

But, to paraphrase T.S. Eliot, between the idea and the reality, there can fall the Shadow.

The cloud that has darkened Sumbly's life – maybe even his soul – began forming some six months ago, when his sister, Nirmala Koul, asked our high commission in India for a tourist visa for herself and her 3-year-old son, Ronok.

The plan, in keeping with the cultural and family traditions of most South Asians, was that she would arrive in Montreal well in time for the birth of the Sumblys' first child and help out with pre- and postnatal care.

That would also thrill little Ronok, who wanted to be there when his cousin arrived. And it would also free Victor to run his pharmacy in the comforting knowledge that Shakti was in good and familiar hands.

Unfortunately, our immigration officials in New Delhi thought otherwise. They refused Victor's sister a visitor's visa for fear that she might claim refugee status after arriving in Canada. The fact that she was leaving behind her husband to run a flourishing business and a 12-year-old daughter to carry on her schooling didn't seem to lend any weight to her and Victor's contention that she had absolutely no intention of abusing her visa conditions.

The visa rejection set off a trail of reaction and outrage that is now into its seventh month and the subject of an investigation by the Canadian Human Rights Commission.

With the same zeal that has become a defining characteristic, Victor is fighting what he sees as a grave injustice to his rights as a law-abiding Canadian.

At issue are answers to seemingly simple questions. Do Canadians have the right to hurdle-free visits from relatives abroad? What happens to this right if the relatives happen to be from countries that generate a large number of immigrants, some of them so eager to migrate that they abuse the system?

Victor's position on these questions is very clear: the system should spot and punish abusers, not penalize law-abiding citizens and their guests.

Even so, while realizing it should not be necessary, Sumbly reinforced his position early in the fight by posting a $50,000 bond guaranteeing his sister's return to India. He filed documents from his accountants testifying to the seven-digit turnover of his business and other assets. He got a bunch of politicians involved, including his MP, Sheila Finestone, and others like David Berger and Lucienne Robillard and MNA Fatima Houda-Pepin. And he got hundreds of his clients and friends to sign a petition and to write to Prime Minister Jean Chrétien.

Faxes flew to and from New Delhi, but over-all, the result of the time and money he spent on his case has been precious little. He's mostly run into walls of silence and evasion. At one stage, immigration officials in New Delhi did relent and offer a visa to his sister, but not to her son. This was clearly unacceptable.

"How could any mother leave her 3-year-old behind, even for a few weeks?" asked Victor in anguish on a recent evening as we talked on his back deck, lamb kebabs emitting tantalizing aromas from the barbecue, his 5-month-old son Dev nestling in his arms.

Indeed.

If there's one lesson he's learned so far, he said, it is that "our politicians don't care about the people. It is not a government run by the people for the people.

"I feel as if I'm living in a banana republic."

Strong and bitter words. But Victor is finally getting some help from an unexpected quarter – Reformer Art Hanger. The Alberta MP, who is known more for his anti-immigration stance, has come out hard and strong in Victor's favor.

Hanger has raised the issue during question period in the Commons. And last week he wrote to Immigration Minister Sergio Marchi demanding an apology for Victor.

"In my opinion," wrote Hanger, "there is no reason that Sumbly's sister should have been denied the visa, even prior to the offer of a bond. With all due respect, Mr. Minister, the only reason that so many requests for visas are denied is because your department, and you, are taking the steps necessary to ensure, that once in Canada, visitors are not able to abuse Canada's overly generous immigration and refugee acceptance system. Once again, law-abiding Canadians and their relatives and guests are paying the price for your unwillingness to take hard action on abuse."

For his part, Marchi, responding to Hanger in the Commons before the latter wrote his letter, told the House that Canada gets a million requests for visas a year from around the world and 85 per cent of them are accepted. However, the minister did not respond specifically to a question from Hanger about Victor's case.

Back again in his pharmacy, sitting now in a cluttered basement office, where we wouldn't be interrupted by Victor's "family" of clients, I asked him what he expected from the case claiming discrimination that he has filed with the federal rights commission.

"Justice and respect for my rights as a Canadian."

And did he feel any differently about this year's Canada Day?

It seemed to take him forever to answer as he gazed at the low ceiling.

"Canada is my country now. I have given my best to this country by giving my service to its people.

"In celebrating Canada Day, I shall pray to God that our politicians become more intelligent and deserving of us.

"And they stop playing games with the people."

Tears, just tears

> How come if you're from England or Europe, relatives have no problems visiting? Are my rights as a Canadian different if I happen to be from Asia or Africa or the Caribbean?

The phone call came in the middle of a busy afternoon of the kind that makes you wish there were two of you. Or even three.

"It's Sumbly," said a voice so flat and muffled that I almost hung up. Things were hectic and I had little patience for unintelligible calls.

But the voice, tiny and muted as it was, also sounded familiar. And there seemed to be something wrong – the caller sounded as if he was crying.

Intrigued, I stayed on the line. Surely this wasn't the same Victor Sumbly I had written about last Monday? The same ebullient Côte des Neiges pharmacist who had been trying unsuccessfully since January to get a visitor's visa for his sister and nephew in India to visit him in Canada? So they could be there for his wife and the birth of their first child?

It had only been 24 hours since he had called to buoyantly thank me for publicizing his battle with Ottawa over his right as a Canadian to host visiting relatives from abroad.

"I had 200 phone calls after your column today," he had proclaimed on the phone. "It's nice to get such support from the community."

I had been inwardly skeptical about the 200 figure but not about the support, because several callers, new as well as native-born Canadians, had also called me and the paper's editor-in-chief.

There had been a curious commonality to these calls. Readers were upset that even a voluntary $50,000 bond from Sumbly wasn't enough to persuade Canadian officials that his sister and her 3-year-old son would return home to India after their visit here.

Several callers had also taken comfort in the fact that Sumbly's problem was a familiar one – they, too, had run into the same uncaring walls.

There was the man in Montreal who couldn't get his father to visit, even though he had done so in the past – suggesting to him a sinister change in attitude in Ottawa.

There was also the man in Ottawa who had wanted his brother-in-law and his family in Ottawa for a wedding. After an initial rejection by immigration officials in Islamabad, Pakistan, the visitors were allowed to come on a minister's permit. When the papers were finally issued, the visitors were given a scant 24 hours to pack and get to the wedding on time at the end of a journey that took up 20 of those hours.

The bitterness and distress, which mirrored Sumbly's, was (and still is) disturbing.

How do you answer questions like: How come if you're from England or Europe, relatives have no problems visiting? Are my rights as a Canadian different if I happen to be from Asia or Africa or the Caribbean?

The first instinct is to counsel patience and understanding. Hundreds of thousands of visitors come here annually, many of them from so-called non-white countries.

Overworked immigration officials, under constant and increasing pressure to keep potential abusers out of Canada, do make mistakes. Legitimate visitors sometimes become innocent victims of their zeal.

But what are we supposed to think when we read that our officials in Jamaica wouldn't let a dead Canadian's sister and son come to Montreal for his funeral? This after the man had died in a freak accident, hit by a truck! Hardly an elaborate immigration scam to bring relatives into Canada.

All these thoughts crowded my mind as I waited for Sumbly to stop sobbing. He could only have bad news.

He did. And it was numbing.

"You know my little nephew in India who was refused a visa? He's dead."

It's a sad, sad twist to a sordid story of bureaucratic and political indifference.

Fed up and stressed out from the strain of struggling for more than six months for the visitor's visas, Sumbly's sister had decided to take a break from it all with her husband, little son and pre-teen daughter.

In the picturesque Kulu-Manali valley that neighbors their home state of Kashmir, little Ronok developed a stomach infection and died of an allergic reaction to medication in hospital.

"Almost the same time you were writing your column about him last week," quavered Sumbly.

"If only they had let him visit me, he would be alive today."

So there you have it. All the ifs that race through your mind when a child dies. All the reasons that could have prevented the death. We all have to die one day, but at the age of 3?

BUNTYS AND PINKIES

In January, Ronok, with all the exuberance of a 3-year-old, had chattered on the phone about seeing his uncle and aunt. And meeting the yet-to-be born cousin he had already christened Dev.

"My sister and Ronok would have stayed here a few months," said Sumbly. "And then my wife, Shakti, and Dev would have gone to Kashmir with them to show our son to our family there.

"He wouldn't have gone to Kulu and he would still be alive.

"We are all devastated."

Would Ronok have avoided death had he come to visit Canada? Sounds reasonable.

Is his death Canada's fault because he was refused a visa? A very murky area, but not for the grief-ridden Sumbly.

But if Ronok's death had further embittered him, he wasn't talking about it. Rather he was thinking of other Canadians caught in visitor's-visa tangles.

"I hope Ronok's death will make the officials more compassionate in their decisions."

Looking to connect

'I have a friend who knows someone who works for the railways," I recall my late father saying one day when he lived in Bombay.

"Why don't you go and see him and he'll help you get a berth on the Rajdhani Express to Delhi?"

"Why don't I go to the train station and get the berth myself?" I remember responding impatiently.

> I call it the Bombay method, but it could just as easily be the Cairo or Lima method. Adrift in a new society, immigrants look for succor among those who preceded them.

Having ridden the rails in scores of cities around the world and flown the skies, too, I deemed it an unwarranted intrusion on this friend of a friend to perform the functions of a reservation counter or travel agent.

Besides, I didn't have the time the process entailed, from making contact with the contact to actually gaining possession of a confirmed reservation on the crack express.

But even so, in the midst of that impatience that my father attributed to living in "foreign lands," it did occur to me that, given the tedious process I went through to book my train my way, perhaps there was more in the friend-of-a-friend method than met the eye.

Today, many years and dozens of train and other journeys later, in the midst of planting roots in one of those foreign lands, in a society rife with touch-tone access to all manner of services, the Bombay method has taken on fresh meaning.

I call it the Bombay method, but it could just as easily be the Cairo, Seoul or Lima method. Or the Dublin or Naples method. The newcomer's network. Adrift in a new society and culture, the world's immigrants look for succor and guidance among those who preceded them.

But the net they fall into is not always the safest. All too often it's an exploitive environment with vengeful undertones. We were here before you and suffered to make things easier for you. But now it's your turn and there's a toll – you have to do some squirming, too.

There are some who are lucky enough to avoid such a net and even luckier (or should I say craftier?) ones who make the net their slave.

However, with time and the understanding it brings, this first net usually becomes a precursor to the next, wider one that eventually expands into an all-embracing one untainted by homogeneity. That great escape, nay release, from the ghetto.

There are exceptions, of course. If you're part of a diaspora, the net can take on different connotations. Forced dispersals, no matter what their scale, leave scars that last through generations and create bonds that outlast many lifetimes.

Diasporas can also spawn tentacles that crisscross the globe. Who can understand the joy of a Sindhi bumping into another one on the south coast of New Zealand's South Island except a third Sindhi?

But then we're talking here – and this is where the filter

becomes personal and unique – of Sindhi-speaking Hindus dislocated by the partition of India. An encounter in Dunedin of this nature would be of little interest to the Sindhi-speaking Muslims and Christians who weren't dislocated from the province of Sind in Pakistan.

Thus one man's diaspora becomes another man's newfound nation until the fault lines of history churn up some new political or other formations.

Doomed, in the meantime, to the present, the scattered people of the diasporas wander the world, ever in search of fault-free havens of peace and stability.

Whether they find one in Dunedin, 20 degrees north of the Antarctic Circle, or here, 20 degrees south of the Arctic Circle, they always keep one ear to the ground, ever afraid that those telltale rumbles from afar might actually be coming from their backyards.

Unsettling and unthinkable as it might be, most diasporics are capable of leaving at the first signs of a crunch. Call them skittish or call them experienced – when you've done it once, the second and subsequent times become that much easier.

Once in the new haven, help and comfort are often a mere phone call away. For people in diasporas, everyone is a friend.

Even if you have to squirm a little.

Moving's a party

I was on one so frequently as a child that to this day a steam train keeps rumbling through my dreams.

Sometimes it's a soothing sound – the train is rolling gently through green fields, the engineer depending on momentum instead of acceleration for speed.

At other times, it's a cocktail of jarring sounds – hisses and skidding steel wheels, the hard shunting of carriages, the coughing gusts of steam.

> 'What, not again! That'll be the seventh address I have for you!' That has become a familiar refrain from my friends, heard most recently when I changed apartments again.

But all the sounds have this in common – the train is always moving, taking me to yet another town, yet another school and yet new neighbors.

Time dilutes the miseries and accentuates the momentary joys of a journey. This is perhaps why I don't remember in minute detail the cramped train trips in overcrowded coaches overheated by the hot Indian sun. Or the biting winter winds that could cut through the thickest of sweaters and the fattest of cotton quilts.

What I choose to recall are the private, first-class bogeys (as they were then called) that my refugee-turned-government-scientist father was entitled to for some of these periodic transfers from one agricultural research institute to another.

One of these bogeys, which had four large berths, armchairs, coffee tables, private toilet with shower and its own doors to the station, would be parked at a siding hours before the scheduled departure of the train it would later be attached to.

So parting for the Chandwanis was usually a long party. The adults would hang around the siding, supervising the loading of trunks that weren't destined for the luggage van, the unrolling of bedrolls, the placement of earthenware water pitchers and other provisioning for a trip that would generally take 24 to 36 hours. (These federally run institutes were scattered all over the vast country.)

The kids would race off to the main platforms to gawk at the milling crowds boarding or alighting from the long-distance expresses and mail trains. And they would gaze wistfully at the fruit and snacks offered by vendors – goodies forbidden to them because of their uncertain hygienic heritage.

Then there were the machines with flashing lights and rotating colored discs that would print your weight on a piece of cardboard at the insertion of a coin.

Had I but collected those weight tickets over the years, I could tell you precisely the number of times we moved. Suffice it to say that it happened often enough to make it a recurring pattern in my life.

Prompted by dislocation and political upheaval, the involuntary journey my parents set out on after the partition of the Indian subcontinent seems to have become one without end, transcending their deaths and the many stops along the way.

"What, not again! That'll be the seventh address I have for you!" has become a familiar refrain I hear from my friends, the last one a month ago when I changed apartments again. I think this was my 27th time since birth, or it could be the 30th.

"How do you do it? I hate moving. I'd go nuts doing it so many times! Doesn't it drive you crazy?" is another familiar refrain.

Does it drive me crazy? It's a question that is starting to bother me, too.

Each time I meet someone who's never left Montreal, I wonder. So, too, when I talk to someone who's only moved two or three times or hear of someone's 70-year-old uncle's first trip out of Saskatchewan.

Who is better off here?

Those who stay in one place all their lives? Children who finish school with the same classmates they entered with?

Or someone like me, who has been through seven different schools in as many cities, four universities in three cities and two countries?

For every 10 parents who dread moving their kids around, worrying about the physical and emotional dislocation, there are at least a couple who see benefits in the future.

New cities, new countries, new faces, new cultures and new schools open up the world. Ask any teacher who might have shipped off to Nigeria or Japan or Indonesia for a year or two with a child in tow. Ask the child years later.

The answers might surprise you with their upbeat message. But sometimes the answers might make you wince, too.

Not everyone travels the same way. Not everyone finds new languages and friends stimulating. Not everyone likes an unending journey.

For some who're always on a journey or involuntarily

stuck in one, that person who has never left home or a home town can become a beacon of stability. And peace.

"Ah, to be anchored!" becomes the secret refrain, drowning out all others.

But then that damn train picks up steam again.

Summer games

Two perspiring men, wearing only shorts, are sitting at a low table on the front porch of a pink bungalow, hunched over a game of chess.

Rising from dead pawns and bishops beside the chessboard are tall glasses of rum and cola, brimming with ice that barely has time to glint in the harsh light of a naked bulb before the heat melts it.

> Ghosts of summers past have a lot to do with the heat when no earthen pot can cool water, rivers run dry and you dream of fans and escapes to lofty mountains.

If you live in the house across the poorly lit street and keep an eye on your neighbors through a chink in the drawn curtains every night, you know that the older man with skin as pink as his house is an ex-pat German priest and the younger, browner and bearded one is a local journalist biding his time to become an ex-pat 10,000 miles away in Canada.

You also know, with that superior contempt only experience seems to justify, that the longtime German neighbor and his temporary boarder from northern India have a lot to learn about coping with summer in Nagpur, a city known for its 45-degree summers.

Look at the fools, you mutter night after steaming night, leaving their windows open all the time, not wearing too many clothes and drinking alcoholic beverages!

Why, you add, any idiot knows that when it's hot you open your windows only at night to bring cooler air in and then shut them all day to keep that air in. And you draw long, thick curtains to keep out the light and the heat it generates.

As for sleeping on the roof, as these two do each night, carrying their bedrolls to the stars, don't they realize it's much cooler to sleep under a fan than on a hot mattress on a hot roof?

Ah, pesky neighbors, whose thoughts always make their way to their target through some gossip grapevine or other. What would the world be without them? Bleak, that's what.

For, sweltering through a summer that for many Montrealers is the hottest in living memory, who else can I turn to for help through the jungle telegraph but the owner of those eyes that my German landlord and I knew were always there?

Bask then, thou ghost from the past, in the knowledge that some of your self-righteousness might have wafted across that dark street and embedded itself in me in some kind of time-release mode.

How else can I explain the numerous arguments I've had had with my new friends about the heat, not just this searing summer but also in the preceding ones?

"Why are your windows closed? Why are the blinds drawn? Why is it so dark in here?" they cry.

Yes, my ghost, you can laugh. Of course you know what I say to them.

But why should they listen? All through the winter they've been behind closed doors, with the skies turning black at 4, the ice and snow chilling covered arms and legs that never feel the wind for months on end.

Who am I tell them to wear cool cottons, cover their heads, stay indoors during the day, eat salt and drink lots of water?

So what if I've been through the same winter and also yearn to splash naked in some cool mountain lake? After all, I'm used to the heat, aren't I, having been born and grown up in it in India?

But they haven't met you, my ghost, have they? Or others of your clan in towns and cities you never knew about. Talk to your buddies and they'll tell you how we slept in airless rooms, dreaming of fans and air conditioners, ice and refrigerators.

Ask them, too, about the water no earthen pots could cool, the rivers that ran dry and the lofty mountains and green valleys only the rich could escape to.

And thus enlightened, my ghost, send out a bulletin on that grapevine. Tell the world that it's just as hard getting used to heat as it is to cold. That the only way to cope with both is to suffer through them.

Tell them, ghost, tell them.

And while you're at it, pull those damned curtains tight, will you?

A facelift for Betty

A poster in the window of Marché Akhavan on Sherbrooke St. in the city's west end advertises a concert of traditional Persian and Kurdish music and songs.

Inside the store, which caters to the culinary needs of a wide range of Asian and Mediterranean groups, you can find pickled or canned vegetables from Turkey or Lebanon, dates from California or Egypt and all kinds of nuts, raisins, beans, lentils, pickles, sweets and breads from around the world.

> In creating a new image of Betty Crocker, General Mills, like other companies, is waking to the fact that we ethnics – don't you hate the word? – have buying power.

Among other products, the shop also sells rugs, curios, samovars and CDs and tapes. And several long shelves are devoted to staples like flour and rice.

On a given day, you might find bewildered-looking customers puzzling over the variety of brands in the rice section, where some of the rice comes in 50-lb. bags.

Here you'll find basmati rice from India with package labels like 444, Real Ricexcitement and Elephant Brand.

Animal figures also grace large bags of scented rice from Thailand with names like Tiger and Jaguar.

Browsing through the store last week, it occurred to me that brand names on display from the home country are so basic and unimaginative that they're amusing, if not laughable. Camel brand tea, for example, which was created for the Middle East market, but is sold at Akhavan, too.

But then, I thought of some of the brand names that have been popular in Canada and the U.S. and decided they too can be a cause of mirth.

I'm talking here of brands like Uncle Ben's, Aunt Jemima, Earl Grey and one that's been in the news lately, Betty Crocker.

But amusing and dubious as the relationship might be between an elephant and long-grained rice from the Himalayan foothills or that between cake mixes and a white, blue-eyed woman, brand names do seem to have the power to provoke fierce loyalties in consumers.

And, as such matters go, the last laugh is usually with the companies and marketing professionals who dream up appealing and profitable brand names.

The overriding principle, whether you're selling laundry detergent to farmers in India or flour mixes to farmers in Saskatchewan, is the same: familiar symbols ensure brand recognition.

Elephants and tigers have been powerful symbols for Asians for centuries, which is why they turn up on everything from soap to rice to pain-balm packages. Betty Crocker's fictional face, one might argue, has been a powerful symbol of middle-class values for millions of Americans and Canadians for the past 75 years.

However, now that the typical middle-class female face no longer has white skin, blue eyes and a "sensible" hairdo, Betty Crocker's masters, General Mills, want to change it.

The company's decision to create a new, computer-assisted image of Betty that would reflect the multiethnic composition of its consumers is hardly motivated by altruism. Ethnic diversity in Canada and the U.S. has been a measurable fact for at least two decades.

No, the real reason is money. Like a growing list of companies, General Mills is waking to the simple fact that we ethnics – don't you hate the word? – have buying power.

To bandy a few statistics about, a study done by T. John Samuel, a Carleton University professor who specializes in these matters, projects that visible minorities will control about a fifth of Canada's GDP by the year 2001, about $311 billion, compared with $76 billion in 1991.

These are staggering sums when you consider that we're not talking of all ethnic groups, just the ones who are not white. The study projects this group to constitute 17.7 per cent of Canada's population in 2001 and compares it to other projections that the group will form 23 to 28 per cent of the U.S. population in 2001.

It follows from these figures that the actual billions of buying power will be much higher when you factor in the incomes of others in the melting pot who may not be from the same background as the traditional, established groups.

Obviously, there's money to be made by broadening Betty's appeal, just as there's money to be made in including blacks, Hispanics, Asians and other minorities in the advertising of other products by other companies.

But in General Mills's case, is the path to multicultural profit a computer-melded image of Betty created from 75 contest-winning photos of real women? Seems to me it is dotted with the same old hurdles.

Altering noses, eyes, hair and skin color on a computer screen or box of cake mix can no more do away with prejudice than clever makeup or plastic surgery.

Companies often argue that they're in business to make

money, not dabble in social change or politics. What they sometimes forget is that consumers, by their actions or passivity, can make dents in such a stand.

Look at Scandinavian stores, for example, which have stopped selling French wine for fear of offending customers opposed to France's nuclear testing.

Those who have money to spend can also choose where to spend it.

Which is my way of saying, I guess, that I'm far more comfortable with a benign-looking elephant on a package than with a politically correct, computer-generated image of "everywoman."

Nightmare for all

> Judges are in the forefront in dealing with the challenges of diversity. If they take a hard line, they're called prejudiced. If they tilt the other way, they're accused of misconception.

It's a nightmarish scene by any standard in any society.

A 20-year-old, pyjama-clad woman runs screaming from her home around midnight to tell neighbors that her father has poured gasoline all over her and threatened to set her ablaze.

Police are called, the man is arrested and charged with attempted murder, assault, death threats and intending to ignite explosive material.

Thus begins a 15-month saga that ultimately results in 51-year-old Popendra Malhotra getting a suspended sentence, 18 months probation and mandatory counseling to deal with his anger, frustration and despair.

Malhotra was convicted of assault with a weapon and uttering death threats but acquitted of charges of attempted murder and aggravated assault. The defence has suggested a conditional discharge because nobody was hurt and no matches or lighters were found at the scene.

Quebec Court Judge Maximilien Polak stuck with the

Crown's recommendation in his sentencing because a presentencing report indicated that Malhotra still denies he committed a criminal act and that he harbors repressed anger and doesn't understand his daughter's behavior or how things work in Canada.

"The family wants to forget the incident and live together again, but that is not realistic under the circumstances," Polak said when he delivered his sentence last Wednesday.

"The aspect of counseling in order to understand a cultural clash (must) be part of the sentencing in this case."

Polak did not elaborate what he meant by a cultural clash, but victim Avanette Malhotra's story, as revealed in court and the media, offers some clues.

The Malhotra family immigrated to Canada in the late 1980s. The gasoline incident was triggered by Malhotra's anger at his adult daughter's "late" arrival at home around 11:40 p.m. on a Friday in June last year.

The incident brought to a violent head tensions in the family over Avanette's decision to terminate a family-arranged marriage that took place in India, to cancel sponsorship of the husband and to pursue a relationship in Canada.

Like most courtroom dramas, this story has its twists. The most puzzling is the behavior and attitude of the victim, who, after her initial terror-driven complaint, claims to have reconciled her differences with her family to the point of being unco-operative with the police and the justice system.

This was the woman, in the words of a police detective quoted in *The Gazette* after the gasoline incident, who "was drenched from head to toe. It was all over the bed and carpet and we called the fire department because it smelled of so much gas."

This was the same woman, in the words of the same detective, who went over to the police station where her father was being held and asked that criminal charges be dropped.

The two "hugged each other, they were crying."

Many months later, after the Crown decided to press charges anyway and the victim did testify on July 24, 1995, Polak found her father guilty.

"As far as the accused was concerned, his body language, it showed he was not telling the truth," Polak said in dismissing claims the victim had doused herself because she was upset.

"He was ill at ease. I could see it, again, through his body language."

Asking for a pre-sentencing report in time for the sentencing last Wednesday, Polak also asked Malhotra and his wife to rise before him after delivering his guilty verdict.

"When you come here from another country with different cultural rules, it is often very difficult to adjust," Polak told the couple and the court.

"And the younger generation, in this case his daughter, may have embraced the Canadian way of life much quicker than he could. But he has to respect that, just as she has to respect her parents."

Polak's remarks in July and last week's reference to a cultural clash highlight the complex web of enlightened motives, myths and perceptions that come into play in our increasingly diverse society.

Judges, far more than police and teachers, are in the forefront of dealing with the challenges of diversity. If they take a hard line, they're accused of prejudice or insensitivity. If they tilt the other way, they're accused of misconception, or worse, of misplaced political correctness.

For Sadeka Siddiqui, co-ordinator of the South Asia Women's Community Centre, to which Polak sentenced Malhotra for counseling, the issues in this case are very clear.

"This is not a cultural thing at all. A crime is a crime. No culture allows you to burn someone if you don't like that person or her views," she said in a telephone interview Friday.

In a separate interview with a *Gazette* reporter, she said: "We don't condone this sort of thing. I think this man should go to prison."

In both interviews, she raised fears that the sentence would encourage violent men to use the "cultural argument" as a defence.

But Siddiqui also pointed out that Polak's sentence was consistent with those generally handed down in domestic-violence cases that do not result in serious bodily harm, a situation women's groups are understandably keen to change.

For her, the case and its issues are not particularly representative of Canada or India. Rather, they're part of a worldwide pattern of violence against women that is the result "of the basic inequality between men and women."

Thus this case and its implications "cannot be restricted to one community – we see people of all backgrounds. Hindu, Muslim, Sikh, Catholic."

Siddiqui's main concern was about the victim and her mother.

"I'm interested in meeting the family. I want to find out more about this man and see if he's controlling the whole family."

So whether Polak's sentence was too light or not, whether or not the daughter is a double victim of the incident and her own fear, one thing is certain.

Counseling for Popendra Malhotra, however angry and confused he might be, is not going to be a picnic.

ASHOK CHANDWANI

Looking for labels

When I heard about Lucien Bouchard's infamous remark about the low birthrate of whites in Quebec, my brown face registered predictable disgust and anger.

But my brain, which arguably is the same color as any other human's, registered fear and sorrow.

The fear welled from the common source shared by minorities in all societies in all countries – it's an unpleasant reality majorities impose, sometimes unintentionally, sometimes with relish.

> Ethnicity and race are not the exclusive domain of separatists. They're also on the agenda elsewhere in Quebec and Canada, whether you're a federalist, a separatist or neither.

The sorrow surged from the fact that it was happening here, in my city, in my country, in my face.

The fact that the remark came from a separatist leader and was followed by an apology did little to mitigate my feelings. For me, to recognize such a context and accept the apology would be tantamount to saying that issues of race and ethnicity are the exclusive domain of separatists.

The truth – and herein lies even greater sorrow – is that

these issues are on the agenda elsewhere in Quebec and Canada, whether you're federalist, separatist or neither.

All Bouchard did was to give public vent to feelings that must clearly be shared by many others. How else can you explain the speed with which the remark has ceased to be an issue in the referendum campaign, with some polls actually indicating a rise in support for Bouchard and the Yes side?

No, to suggest that race is an issue only for separatists, sovereignists, nationalists or whatever label you want to give them is to ignore that it also affects everyone else.

Why, it was only last month that Ottawa decided without much fanfare to add a question about race to the 1996 census.

Until now the census used to extrapolate on the demographic mix of Canada from answers to ethnocultural questions about respondents' ancestry and language.

Now one in five Canadians who will get the long response form will be asked to place their race in one of 11 categories – white, Chinese, South Asian, black, Arab/West Asian, Filipino, South East Asian, Latin American, Japanese, Korean or other.

Ottawa has said that this question is necessary to obtain information considered vital to administer equal-opportunity and employment-equity laws.

Opponents of Question 19, who include academics and readers writing letters to newspapers, argue that it invades privacy and offers a dated concept of race.

Indeed, if you pause to think about it, how would you define the term "white race" as required by the census? (Remember, as you struggle for an answer, Bouchard's reference to it and also that it is illegal to not answer census questions, with fines ranging up to $100.)

Arguably, the only sensible course left to anyone unwilling to be labelled by ethnicity would be to enter the word "human" under the "other" category. Or, perhaps, "Canadian."

Apart from infuriating the number crunchers at Statistics Canada, such an answer would also create an intriguing new scapegoat for politicians and lobby groups.

Imagine how much more difficult it would become for Quebec separatists to suggest – as some have been doing lately – that allophones are planning to vote No as the result of some secret and undocumented conspiracy to undermine a legitimate sovereignist movement.

Imagine aiming such rumblings at Quebec's "human" or "Canadian" population?

Just wouldn't wash, would it? For to do that would be to confer on allophones a mantle of inclusivity and equality that is just not in the interest of the majority.

Alas, the census is not until next year and there is no way to test its results.

For now, I guess I just have to cope with all my fears and sorrows.

This, too, shall pass.

That sinking feeling

"It's like when the jury is out," says a friend. "There's nothing to do but wait."

Others gathered at a table in the ballroom of the Ritz-Carlton Kempinski hotel for a literary event called Books and Breakfast nod sombrely.

Conversation among the casually dressed book-lovers at other tables appears hushed.

> For weeks I have been wrestling with today's referendum, my nightmares and dreams colliding with those churned up by the Yes and No campaigns.

The ornate hall is ablaze with chandeliers, but few seem particularly bright or cheerful as they tuck into their mushroom scrambled eggs and croissants, waiting for the invited authors to speak.

A sense of weary dread seems to permeate the air as Michael Goldbloom, publisher of *The Gazette* (which sponsored the event with the Paragraphe bookstore) and city columnist Peggy Curran welcome guests and introduce the writers.

All these years, quips Goldbloom in passing, we've all enjoyed the extra hour we gain on the last Sunday of October. But this year it's an extra hour of "water torture."

People laugh nervously, almost reluctantly.

The same laughter greets Curran when she talks about the "fur ball" in her stomach and a party the previous night where everyone "danced in a frenzy" for four hours and then didn't know what to do next to avoid thinking of the results of today's vote.

It's bracing, this collective desire and desperate ability to laugh and dance away the knots in our stomachs. But it does little to blunt the chill in my heart, even though I can't help but laugh along with the rest.

It's hard to explain this chill, to confront this gnawing fear that has been building as ominously as the trend lines in the polls.

The writers talk about their books. Linda Leith tells us about a novel set in Pointe Claire about a tenant from hell. Ken Dryden discusses a book that looks at the school system. Ron Atkey describes a book about political intrigue and Canada's security and intelligence apparatus.

It all sounds so normal. Just as it does when they talk about their close, often inextricable, ties to Montreal.

Yet, inside me, nothing seems normal. I can't concentrate. For weeks I have been wrestling with today's referendum, my nightmares and dreams colliding with those churned up by the Yes and No campaigns.

Which nightmare should I ignore? Which dream should I buy or cling to?

The worst (and recurring) nightmare I have is that I would become a refugee in Canada in the aftermath of Quebec separation.

Before you scoff or laugh, remember it's a nightmare, so it's not particularly rational. Remember too, that people like me – one of 5 million foreign-born Canadians – each have baggage from the past that we try conscientiously to suppress, but which lives a life of its own.

My baggage is that of being the child of a refugee family dislocated by the creation of the new state of Pakistan in 1947 by the partition of India. Yes, the circumstances were different, the motivations were different and the history was different. But the net result was that, to remain Indian, my family had to become destitute refugees in India.

Compared with this personal nightmare, the spectre of a reduced standard of living in a free and democratic Quebec doesn't sound that intimidating. Except, when I think it through, it clearly means that I would have to give up on being a Canadian.

This numbing realization triggers a fresh conflict, which I dare say might strike a chord in other foreign-born or native allophones: what should motivate federalist allophones – pragmatism or patriotism?

Pragmatism dictates that, given Canada and Quebec's history of accommodation and compromise – a history missing in the blood-ridden scenarios of Indian partition and the Balkans – allophones would find an acceptable niche in a new Quebec state.

Patriotism dictates that love for and commitment to Canada overrides pragmatic reasons for remaining in Quebec, no matter what the personal toll of dislocation.

It's an unsettling choice – the kind that leads to nightmares. But when you live in interesting times, you have to confront your demons, take a stand and live with the consequences.

I know I'll be able to survive with my decision to vote for Canada by voting No today.

I just wish, as I sip my tepid tea at the Ritz, that the verdict from that massive jury we're all waiting to hear from will bolster mine.

Room for all

> Jacques Parizeau's deplorable remarks about ethnics have led to a remarkable pan-ethnic bonding. This is a tiny victory for the forces of cosmopolitanism.

As denunciation and soul-searching continue in the aftermath of last week's referendum, take a moment to look at the bright side.

OK, OK, not a whole side, but a sliver. A really thin one that's there if you care to step aside from the rhetoric.

Sure, we've all emerged divided and bitter. Yes, everything that Jacques Parizeau said about money and ethnics sounds as "sinister" a week later and is as "reprehensible" and implies a "form of restrained ethnic cleansing," to use the words of people and politicians quoted in *The Gazette*.

But in all the tribal finger-pointing and bleating that has been heard from all corners of the debate – and indeed the country – surely some of us can claim a tiny, tiny victory for cosmopolitanism.

How else can you explain the remarkable pan-ethnic bonding that has followed Parizeau's deplorable remarks and the dubious behavior and statements of some of his supporters?

While francophones, anglophones and allophones have all heaped scorn on the negative ethnocentric implications of Parizeau's speech, quite a few from the last two groups seem to have found far more common ground than the voting conspiracies and inflexibilities they have been accused of.

When people of Chinese, Caribbean or Latin American heritage talk about certain commonalities of interest with someone from the Indian subcontinent, it's reasonable to assume that there's a shared thread and history of experience linked to skin color.

But when they're joined by people of Irish, Ukrainian, English, German, Scottish and Italian background, then it becomes a different story altogether.

To me, this type of cross-ethnic bonding, however temporary and however tongue-in-cheek, and even if it is in the name of a larger tribe called Canada, is a welcome portent for the future.

When a German Canadian shouts across the bar at me that "finally we're being recognized," he's doing much more than trumpeting his joy and relief that Canada emerged intact from the referendum.

No, to me, he's saying what a clear majority of Montrealers said on referendum night – that, when it comes to the crunch, a vote against the separation of Quebec is not just a vote for Canada, it's also a vote for cosmopolitanism.

Call me naive and lunge for my rose-colored glasses, but he – they, we, pick your pronoun – is also saying that tolerant, free and self-confident societies can best be nurtured through cultural and ethnic diversity.

Naturally, given the tumult, turmoil and cruel violence that has shaped the world and its nations, cosmopolitanism faces a great many enemies, chief among them the forces of xenophobia and tribalism.

We've seen these forces take their deadly toll in recent

years in the Balkans. We've seen them active in the anti-immigrant violence in Germany, France and Italy. We've seen them shatter the peace among Hindus and Muslims in India. And we have heard their echoes on this continent on extremist hot lines and talk shows.

This is primarily because any struggle between cosmopolitanism and its enemies is an unequal one – the latter clearly have superior motivation, morale and, in the theatres where violence is commonplace, heavy weapons and firing power.

The only weapons cosmopolitanism can boast are those of inclusivity, mutual respect, tolerance and acceptance. Which pretty much means that cosmopolitanism has a better chance of thriving in the kind of peaceful, democratic setting we're wedded to in Quebec and in Canada.

Where else could we mark ballots, raise voices and spout words that question the belief that cultural identities are homogeneous and impermeable and can only be preserved in isolation of all external influences?

However, for cosmopolitanism to grow and plant roots, its proponents will have to look beyond the issues that so divided people last week.

By definition, tribalism afflicts all people and all groups.

The challenge of that sliver-glimmer is to find room and respect for each of us.

Brutal truths

Assia Kada's eyes cloud, reflecting the pain in her words. The fingers of one hand define space the size of a tennis ball around one eye, as her friend Liza Novak nods in sympathy.

"I have seen," says Kada, "a woman with her eye swollen like this, with blood hemorrhaging from her ears.

> Immigrant women are usually unfamiliar with local laws and individual rights. This makes it difficult for them to escape from the violent men in their lives.

"I have seen women's faces deformed from being hit.

"I have seen women with their arms and legs and noses broken.

"I have seen fractures. I have seen bruises and I have seen burns – cigarette burns."

Kada's voice is even and controlled, the pain restrained to counterpoint the inherent brutality of her words. Kada is not speaking with intent to shock. No, she, like Novak, is merely offering a chilling peek into the world of Secours aux Femmes, a shelter that deals primarily with the problem of conjugal violence encountered by immigrant women.

This week marks the sixth anniversary of the Polytechnique massacre that claimed the lives of 14 women, most of them engineering students. It's a good, if sad, week, Novak and Kada say, to focus on the unabating violence against women and reach out to those who might be unaware of the existence of their modest 12-year-old shelter.

Men abuse women in all cultures and all groups, the two acknowledge. But Novak, whose parents came here from Poland, and Kada, who came here from Algeria, point out that immigrant victims of violence have special needs that require special services.

More often than not, there are children involved. Of the 210 victims who took refuge in their shelter last year, 100 were women, the rest children. This means most of the abused women Secours aux Femmes gave shelter to were or ended up being single mothers. On a given day, six or seven of the shelter's 15 beds are occupied by adults, the rest by children, some of whom might be infants.

Immigrant women, especially newly arrived ones, are often limited in French and English and are usually unfamiliar with local laws and their individual rights. All these factors make it harder for them to escape from the violent men in their lives and pose special challenges for shelters like Secours aux Femmes, whose doors are also open to all abused women.

While the Quebec government, through its health and other agencies, bears the operating cost of maintaining 15 beds and a small staff of six, like other shelters, Secours aux Femmes is forced to seek and depend on donations for its other services.

As Kada and Novak point out, a woman's problems do not end the moment she makes the momentous decision to put a stop to the violence against her by leaving her abuser for good. This decision, which rarely results from the woman's first refuge in the shelter, leads to new and unavoidable problems.

The victim needs help in finding a place to live; to obtain welfare, if she doesn't have a job; she needs someone to accompany her for protection as well as translation for legal actions and proceedings; she and her children need assistance to buy food, clothes and toys.

The list is endless but essential, and the only funding and time that go into these activities is voluntary. No shelter worker anywhere works office or regulation hours.

Novak, who is the co-ordinator at Secours aux Femmes, and Nada, who is a caseworker and intervenor, are no exception.

The long hours they put in give them a unique insight into the nature and patterns of violence against women, which is far more common than many might think. Statistics indicate that more than one in four Canadian women, regardless of background or income bracket, are victims of abuse, be it verbal, emotional, physical or sexual.

The pattern of violence against immigrant women, Kada and Novak say, sometimes reflects the shifts in balance of source countries. These days, their caseload reflects a peculiar and disturbing trend affecting immigrant women from Latin America, Mexico, Haiti and the Dominican Republic.

The shelter has seen a dramatic rise – last year they had about 45 cases – in the number of abused teenage brides brought here on spousal visas by Quebecers. They've seen victims as young as 13 but usually around 18 who married Quebec men in their late 50s. One case involved a man who was 60!

Driven or enticed by dreams and promises of a new, poverty-free life, the teenagers, who arrive on perfectly legal sponsorships, find themselves trapped in the clutches of abusive and violent men.

"He told her he had a big house and two cars," says Kada of one case in which a battered bride took refuge. "But all he had was a three-and-half (room) apartment and a bicycle!"

Novak and Kada both ask, with justifiable anger, whether immigration officials ever take a closer look at spousal-visa applications that involve "grandfathers" and teenagers.

And they are at pains to debunk the "myth of culture shock" that "society and the media" use to explain or condone the violent behavior of immigrant men toward their spouses.

"It has nothing to do with culture or the shock of being in a new country," says Novak.

"Nor has it do with alcohol, loss of a job or socioeconomic status. We've had cases where the man earned more than $100,000 a year.

"No, for us (at Secours aux Femmes) it's an issue of control and power.

"When a man's control and power are threatened, it triggers conjugal violence."

Pangs of bondage

R aju was 11 when I first met him, a doe-eyed wisp of a boy, hired as a domestic helper by my parents shortly before my father died in May 1988.

Raju was there in the kitchen of a small two-bedroom flat full of mourners in Bombay when I arrived after a 36-hour scramble for visas and a never-ending flight for my father's funeral rites.

> Raju was too young and too smart to be working. There was also an air of mystery about him when he wasn't flashing his winsome grin or deftly trimming okra.

He was brewing tea by the gallon for the mourners gathered in a wake-like atmosphere around the flat, many of them clustered around my mother, confined to bed by a bungled hip operation.

I was glad to see Raju – for months I had been urging my brother to hire a cook-helper to ease the burden of heart attacks and hip injuries on our enfeebled parents.

But his age and deportment made me uneasy. Raju was too young and too smart to be working.

There was also an air of mystery and inner sorrow about him when he wasn't flashing his winsome grin or deftly trimming okra for the evening meal.

For the next two weeks, as the mourners ebbed and flowed and eventually dried up, Raju took over my mind as much as my departed father.

The death of a parent inevitably stokes memories of childhood. As the child of parents forced by the partition of India to become refugees, I had seen and experienced many of the ugly facets of poverty.

But as I watched Raju perform his chores, I was acutely reminded that my parents' deprivation had not forced them to send me off to some distant metropolis to work as a child domestic.

For this is what had happened to Raju, as much as I could glean from the limited information available. All my mother and brother knew about him was that he had come highly recommended and been brought to them by an uncle of Raju's from his "muluk"– home county.

This uncle was part of a family of male migrants working as domestic helpers in Bombay. He collected Raju's and other relatives' wages, remitting the bulk of them to their muluk to wives, mothers, daughters and other dependents.

Once a year, they all took a month off to spend in their muluk.

It was a muluk – actually a village near India's border with Nepal – that Raju rarely talked about, affecting a vagueness imposed by the vagaries of servitude.

Do your "duty" honestly and keep your employers happy, Raju had been brainwashed by his uncle, but don't talk about your muluk, lest they or the police ship you back there.

The few occasions I was able to coax Raju into talking about his muluk, he would paint a fanciful picture of an idyllic place in the mountains, full of happy families and lots of plump goats.

Whether Raju's words were intended to soothe my guilt

or not, they always failed to do so. And yet I felt totally powerless to do much more than brood about Raju's lot, not the least because after years of living in Canada, I felt somewhat foreign in the country of my birth.

Who was I, I kept telling myself, to question my family's domestic arrangements? Having chosen to leave, I felt, had deprived me of any say in the matter.

Besides, I was there for a mere three weeks.

Yet it was impossible to ignore Raju's plight, in spite of its relatively benign setting. But three square meals, clothing, shelter, wages and watching television aside, there was no denying that Raju was child labor and illegally so.

Still, as my family and friends argued, turning Raju over to the cops or even to his uncle would not solve the problem. The cops would put him in an overcrowded and underfunded home where he would be worse off. Or, out of pragmatism, deliver him to his guardian and uncle, who would put him in a job with some other family.

The more I brooded, the more it became clear that there was no clear or handy solution to this dilemma, fueled as it was by the inherent contradictions and conflicts of a developing country.

I ended up consoling (deluding?) myself that at least Raju was somewhat better off than the 50 million Indian children forced by poverty to work under exploitive conditions in carpet-weaving and other labor-intensive industries.

These practices and conditions, illegal and reprehensible as they are, show little sign of abating, in spite of the best intentions of legislators and the protests of human-rights activists, the latest among them being Craig Kielburger, a 13-year-old globe-trotting schoolboy from Thornhill, Ont.

It was watching and reading about the fresh-faced Boy Scout's crusading energy and flashing eyes in New Delhi last week that resurrected Raju from those funeral days in Bombay and all the entrenched dilemmas.

Once again, I wondered what ultimately became of that little charmer with so much hidden pain.

For, one day, not that long after I returned to Montreal to resume a life that was now fatherless, Raju secretly packed his little tin suitcase and slipped out of the Bombay flat, never to be seen or heard from again.

A flag's just a flag

As the last scene faded from the movie screen, I jumped to my feet to stand at attention for a clip of the Indian tricolor and national anthem.

I was 16 then and on a first furtive date at a movie in New Delhi that had an 18-plus restriction. Clearly, it was important that I behave as a cool grownup for the sake of my friend and the suspicious ushers.

Thus, rising with alacrity for the anthem, a patriotic screening decreed by the Indian government in the wake of wars with China and Pakistan in the 1960s, seemed a good way to send a message that here was a savvy guy who knew his way around.

Alas, something went grievously wrong.

Instead of the national flag and anthem, I found myself staring at one more scene from the movie that I thought had just ended.

I also realized I was the only person in the audience who

> In an era of globalization, the commendable attitude should continue to be one of quiet pride in one's country and heritage and a healthy respect for that of others.

was standing. Embarrassed as only teenagers can be, I sat down as fast as I had risen, forgetting that my seat had sprung back when I stood up.

The physical bruises of crashing ignominiously at the feet of my date have long healed. But I have yet to erase the memory of her unrestrained laughter and that of people around.

Still, in hindsight, I feel some debt to those guffaws, because they made me question, for the first time, the concepts of flag and patriotism.

Until that day of misdirected fury at the national flag – the cause of my downfall – I had been content to blindly sing, march, salute, stand at attention or perform other internationally sanctioned acts of respect for flag, anthem and nation.

The movie incident initiated a change in attitude that was cemented, subsequently, by the worldwide ferment that gripped youth in the mid-'60s through the early '70s and the movements for world peace.

Immigration and a change in citizenship and allegiance contributed immensely, too, to delinking pride in country and heritage from jingoistic chest-thumping about national symbols.

Over more than two decades in Canada, it always has been comforting to see that, for the most part, we don't make a big deal about flags.

The self-deprecation and understatement that are often mentioned as Canadian characteristics have always been apparent in the tiny size and discreet use of flag pins.

In an era of increasing globalization, the commendable, cosmopolitan attitude has been one of a quiet pride in one's country and heritage, accompanied by a healthy respect for that of others.

This is an attitude that has endured in spite of the flag-filled parades, marches and rallies of recent years, many of them linked to the never-ending unity debate.

But as we mark Heritage Day today, this attitude seems to be under scrutiny and, some might say, attack.

Last Monday, at the beginning of Citizenship Week, Ottawa proclaimed Feb. 15 as Flag Day to mark the 31st anniversary of the adoption of the Maple Leaf as our national flag.

The result wasn't quite the public outpouring of pride and patriotism the federal government would presumably like.

Rather, on the day itself, during ceremonies marking it in Hull, Prime Minister Jean Chrétien found himself embroiled in an undignified and deplorable scuffle with a political protester and heckler.

Whatever the fallout from the incident, it certainly is another little portent in a growing series that collectively seems to signal a departure from the peaceful and peaceable attitudes to which we've all become accustomed.

I don't know how eagerly people will respond to suggestions that we all proclaim our patriotism more visibly and actively.

But I do know that when I tried to do so in a different time and place, it left many an unwelcome scar.

A stranger to all

> A stray remark at a cultural event, a movie set in the Old Country, a whiff of certain food aromas – any one of these can make you yearn for the place you've left behind.

Sitting on a sea wall near the Gateway to India in downtown Bombay, my Canadian travelmate, reading aloud from a phrasebook, said loudly in phonetically correct Hindi:

"Aap chaleh jao, budmash!"

The words – "go away, you ruffian!" – made one of the scores of Bombayites enjoying the evening breeze around us start and look at us in suspicion.

Oh oh, I thought, here's trouble. He's going to think I'm harassing this foreign woman and upbraid me for it.

Instead, to my surprise, the man asked in English: "Do you know what those words mean?"

It was clear that he thought I, too, was a foreigner and that we were playing with words we were both unfamiliar with.

When I answered in Hindi that yes, I certainly knew the meaning of the words, it was his turn to be surprised.

"Oh, I didn't realize you were Indian," he replied, sticking to English for the benefit of my friend.

That encounter and a resulting short and convoluted conversation about nationality took place seven years ago. But the incident keeps popping up each time I experience an urge to return to the Old Country. Not for a holiday, but for good.

It's an unsettling and mysterious impulse, this, fraught with implications, some enticing, others disturbing.

I don't know how many of nearly 5 million foreign-born Canadians experience this urge. But I suspect it's a significant number, particularly among those who arrived here as adults.

The usual trigger for such an impulse, I have found, is nostalgia. A stray remark, a cultural event, a movie set in the Old Country, a whiff of certain food aromas – any one of these can make you yearn for the country you've left behind.

These nostalgic twinges are rarely more than momentary. When they become ulcerous, a holiday to the Old Country, if affordable, seems to be the best cure, though not necessarily a completely pleasant one. A trip to the Old Country, returnees discover, can also confront you with the reasons you left in the first place.

For those unable or unwilling to make the trip, there are affordable modern alternatives – the telephone, satellite television and videos.

There are cheaper antidotes, but these can prove costly in other ways. The perils of withdrawal – whether to a darkened room or a ghetto – are well known.

Still, nostalgia is not the only kindling that turns newcomers' thoughts to their former haunts.

There is plenty of other fuel, some of it flowing from the Old Country. A political regime might change; the economy might be booming. Conversely, things might get dicey politically in the new country, the economy might stagnate.

Personal experience plays its own important role. A newcomer might be unable to adjust to his new country. He might face racism, job discrimination or underemployment. A marriage might fail or a romance end in rejection. Friends might be hard to find or keep.

Whatever the spark, the urge to return can only create turmoil in the victim's mind. The turmoil can flare into agony when there is no one reason, but a cocktail of them.

And as much as the presence or lack of parents, spouses, children, lovers and friends can flavor the cocktail, the antics of politicians can poison it.

Suddenly, you're forced to examine all your assumptions about country and identity.

What does home mean? Is it here, back there or somewhere else?

What does nationality stem from, birth or a passport?

Where do language, religion, culture, customs and ethnicity fit in?

Who belongs where? Who accepts or rejects whom?

Allegiances blur, loyalties clash.

You become a stranger in the new country – and the old.

The man in Bombay had as much trouble seeing me as an Indian as many here have in seeing me as Canadian. It couldn't have been my looks – there were plenty of people on the sea wall with similar features and coloring. Was it some association created by my travel-mate? Speech patterns or accent I had acquired unconsciously? Gestures, body language or clothing?

I don't know, because a sudden downpour ended the seawall encounter. But even if it hadn't rained, I doubt there would have been any answers. Not satisfying ones, anyway.

The questions raised by the return urge are far too complicated to be resolved on sidewalks or, for that matter, in any one place at any one time.

They lurk, like ruffians everywhere, to attack you unawares. All I can do is shout "chaleh jao!" and hope, in vain, that they obey.

As the wheels turn

In the early 1960s, one of those Hindi musical fantasies Bombay's film industry is known for churned up a song that became such a hit its lyrics still jingle in my mind.

Loosely translated, the opening of the song went something like this:
> *I'm a rickshaw-wallah*
> *I'm a rickshaw-wallah*
> *This man with two legs*
> *Is equal to (one) with four*
> *Where can I take you, babu?*
> *Where can I take you, lala?*
> *I'm a rickshaw-wallah*
> *I'm a rickshaw-wallah*

> The rickshaw, long a symbol of poverty and exploitation in large chunks of Asia, is being touted as an environmentally friendly form of transit in Canada and England.

It was the catchy music of the song and the star who lip-synched it on screen that made it a hit, not its inherent poignancy.

The miserable working conditions of rickshaw-wallahs were lost on their passengers, whether they were the office clerks, known as babus, or fat-cat merchants, called lalas.

All manner of people still use rickshaws in many Indian cities and towns. Schoolchildren in smart uniforms ride the two-passenger vehicle in groups of four to six; the babus ride them with their families and luggage to and from railway stations; the fat lalas and their equally fat wives ride them to the movies.

But the typical rickshaw-wallah remains an impoverished, emaciated slum dweller coping with a truncated life expectancy because of respiratory disease, heart enlargement and arthritis.

These conditions have hardly changed since the rickshaw first appeared in 1870 in Japan as the jinriksha, a hand-pulled, two-wheeled contraption with a passenger seat and folding hood invented by a missionary named W. Noble.

Today, most of the hand-pulled rickshaws in the Indian subcontinent, China and southeast Asia have been replaced by three-wheeled cycle-rickshaws, sometimes called pedicabs, or by motorized or "auto" rickshaws.

The pedicabs are still powered by human legs – instead of running, the driver pedals. The autos, also three-wheelers, have scooter engines and gears.

But even as the rickshaw has evolved and survived agonized discussion about the plight and fate of the millions of pullers and pedallers in its Asian habitat, it has also migrated to Western streets.

Many Canadians have seen and used high-tech tourist rickshaws on downtown streets in Montreal, Vancouver, Toronto and Ottawa, although only the latter two cities appear to have them with any consistency.

In all these cities, but particularly in Toronto, the rickshaw, generally driven in the summer by fit young male students, has caused controversy. However, the controversy has centred on regulation, licensing and cases of price-gouging.

That doesn't appear to be the case in Cambridge, England, where the rickshaw is being introduced in a one-year trial as a solution to the historic city's pollution problems.

According to a report in the London *Observer*, the 500-year-old chapel at King's College in the heart of the city has suffered $4 million worth of pollution, partly from traffic fumes.

Rickshaws, the city's environmental committee reasons, would not only cut down on the fumes but offer transportation to the 350,000 people who visit the chapel every year.

The basic vehicles will be imported from India and adapted for local use. The seats will be widened "for Western bottoms" and the British-made chassis will have 21 gears, hydraulic brakes, halogen lights and superior suspensions.

"The result is that a woman of average build and fitness can cycle them," an official with the company bidding for the rickshaw contract is quoted as saying.

To which, all I can say is wow! – more in irony than in admiration. An invention that has become a symbol of poverty and exploitation in large chunks of Asia is now being considered environmentally friendly.

Whereas Asian governments struggle to replace human-powered rickshaws with fume-causing autos, in Cambridge they want to use them to control pollution!

About the only common denominator in this absurd divide is economics: in Asia as well as among students in Canada and Britain, the rickshaw offers crucial income, be it for a livelihood or university fees.

And, if the recession-driven experience of Toronto is a guide, jobless workers aren't averse to driving a rickshaw for a seasonal livelihood.

This twist in the rickshaw's fortunes might make some of us laugh, but for the tens of thousands of Asians eking out a living from rickshaws, it's no joke.

They remain doomed to a life in which an occasional song of fake bravado offers momentary cheer.

What price tradition?

> The ill-defined 'family values' that right-wingers talk about seem to be forcing their way into the consciousness of baby boomers who pride themselves on their liberalism.

An animated dinner conversation about families dredged up an image of our departed mother spoon-feeding my younger brother.

What was remarkable about this normal maternal practice was the age of my brother at the time – 28.

There he was, dressing for work, putting on a tie, combing his hair, while my mother followed him around their Bombay flat with a plate of breakfast, helping him eat on the run.

Equally remarkable was the doting love in her demeanor and my brother's nonchalant acquiescence.

Whether it was because of my living alone and away from home since the age of 15 or my different life experience, I found their behavior disconcerting.

It wasn't that I felt deprived in any way during my years at home or away. It wasn't at all a case of imbalanced affection on the part of our mother or jealousy on my part.

No, it was simply that I could not comprehend adults in

such a situation, choosing to see it as something that ought to peter out as children move on from teenhood.

My muted protestations (I was on a short visit from Canada and didn't wish to provoke a family squabble) fell on amused ears.

"You don't know how a parent's mind works because you don't have children," my mother replied. "She likes doing it, makes her happy," my brother said.

Throwing this 10-year-old episode into the dinner discussion over the weekend uncorked a chain of reminiscences, confessions and recriminations in the mixed group of Italian, Lebanese, francophone and anglophone background.

The trigger had been a story about a couple moving in with in-laws while their home was under renovation. But the talk widened to cover breast-feeding, children sharing parental beds, allergies and gender-linked domestic chores.

With everyone at the table a few years on both sides of 40, it struck me that our concerns echoed other such conversations I've had recently, and are perhaps a barometer of baby boomers' disillusionments.

Having achieved material comfort but faced with the vicissitudes of a downsizing, cost-cutting economy, they're discovering disturbing gaps and contradictions in their lives.

Those ill-defined "family values" that many older conservatives and right-wingers talk about seem to be forcing their way into the consciousness of boomers who pride themselves on their liberalism.

Not surprisingly, the discourse is influenced by the increasing diversity of our society, with immigrants or their children seeking succor or explanations in the value and belief systems of their particular heritages.

Thus you hear people saying things like "Our children would never be with babysitters back home" or "You never argue with a Sicilian father" or "Extended families are best for kids." Whatever the validity of such responses (there are arguments in favor of babysitters and against extended families), one thing is certain – people's quest for answers is genuine.

Ironically, the expanded availability of information – this is an age when long-lost children can track a parent on the Internet – can add to the confusion. Often a new study seems to contradict a previous one. Look at the issues of breast-feeding or babies sleeping with their mothers.

For long centuries, these were (and still are) normal practices around the world.

But in recent decades, bottled formula and cribs became the norm in North America – practices that finally seem to be reverting to the traditional ones, though not without misinformed resistance.

The technologies of information and communication also appear to be taking their toll on other relationships. While the telephone, television and Internet all boast of (and do have) tremendous power to educate and unite people, all too often they also breed isolation and impersonal contact.

Is the answer, then, some regression to a hoary past when families stuck together and neighbors and relatives found time to visit? Even if it were possible to re-create this rather romanticized past, my answer would be No.

This is not because I adore my cell phone, surf the Net or spend countless hours glued to a TV set. Far from it.

No, it's because I believe these and other technological marvels can be harnessed to our needs and that we can find acceptable answers in spite of them.

Seriously, whatever the family environment of a child, who would want to give up on the benefits, say, of an occasional carefully chosen video?

At the very least, it would open a window of communal time for people to tackle larger issues like in-laws' attitudes or moms who spoon-feed 28-year-olds.

Choo-choo magic

> The train is a personal and universal metaphor for migration. The journeys of my uncle, my parents and millions of others have rarely ended after that first train ride.

Driving to work in the morning, a colleague and I deliberately follow a meandering route that crosses railway tracks just north of Notre Dame St. in southwest Montreal.

Long before the intersection appears, my heartbeat quickens. Will it happen today? Will we, oh, will we see a train?

When we do – it's always unexpectedly because our driving times and rail times keep changing – the stars seem to shine by day, renewing my lifelong romance with trains.

It doesn't matter whether the train is a sleek, yellow-blue Via passenger bound for Toronto or a lumbering long freight carrying cars and cows.

The thrill is the same one I have felt since childhood in the eastern Indian city of Cuttack whenever a train held up our school bus.

Others fretted, but I kept my nose glued to the window, gaping bug-eyed as the Madras Express flashed by or a steam locomotive marshalled freight cars.

Those trains and their sounds – whistles and hisses, the clanging of iron – have been with me all my life on a journey that has spanned three continents and 50 countries.

Sometimes the journey has been as exotic and glamorous as those we see in the movies – From Russia With Love, Dr. Zhivago and the latest, Bullet to Beijing by Montrealer George Mihalka, which hit theatres last week.

No suave ruffians with guns have burst into my compartment and no sultry seductress has begged for sanctuary on my exotic trips. But the clackety-clack magic of the wheels has always been the same in the wagon-lit sleeper from Frankfurt to Paris, the Trans Europe Express from Zurich to Rome, the Rajdhani Express from Bombay to Delhi and the grand salon of the Via Transcontinental from Montreal to Vancouver.

On the marathon 72-hour run to Vancouver in 1986, a kindly service manager fulfilled a lifelong dream – he let me and a friend ride in the locomotive from Dryden to Kenora, Ont.

There I was pulling 17 coaches, barrelling along at 70 miles per hour through central Ontario, blowing the horn at every road crossing – two long toots and a short followed by a final long.

The last time I had been so close to a tooting horn had been as a kid on summer holiday in a Bombay suburb where an uncle was an assistant station master. The courtyard of his modest, railway-assigned two-room house was a scant 30 feet from the tracks, just where the southbound platform of his station sloped to an end.

Electric commuter trains stopped and resumed their mad dash to downtown every few minutes for 22 hours a day, tooting their horns literally in our ears. And all day and all night, the long-distance expresses and freights thundered by. It was the stuff of nightmares – or dreams. It being summer, we slept outdoors in the courtyard, which vibrated constantly. But the nights were cool, there were stars in the sky and sometimes a moon. When the trains stopped running between 2 and 4 a.m., it was the silence that jarred.

It was a silence not unlike that on board a train, when, after hours of rhythmic rolling, it stops in the middle of the night in the middle of nowhere. You look out and see shadows flit ominously in the dim light of signal lanterns. For some, the scene adds to the magic. For others, it spells menace.

Romantic as it might be, the train is also a grim symbol of suffering for countless millions of people. It is the train that has taken millions to their cruel deaths or to labor and concentration camps; it is also the train that has seen ethnic and religious bloodbaths, often on board, even as it is being used to help people escape.

A few years before I was born, it was a train that brought my parents to what was still India after its partition in 1947. They were helped in their flight by an uncle who made several return trips across the violence-ridden border to escort my aunts and grandparents to safety.

Not surprisingly the train has become a personal and universal metaphor for migration. For the journeys of my uncle and my parents and millions of others around the world have rarely ended after that first train ride.

Inevitably, there have been more journeys, voluntary and involuntary, as people seek better lives or flee upheavals.

If, as an adult, I have sipped wine and slept on linen-covered berths from Calcutta to Kenora, as a teenager I have also slept under a wooden berth, the only free space in a crowded third-class coach.

On all these journeys, cramped or in luxury, amidst the rhythm and the din, the joy and the pain, the one blissful constant that has emerged is that there is no one constant. Just endless rocking and rattling, an undying quest for what lies at the end of the line. And when you do arrive, it's only to pause briefly and start again.

Confession time

A couple of weeks ago, I received the long version of the census form mailed to every fifth Canadian.

This is the form that contained Question 19, which offered Canadians 11 categories to specify whether they are white, Chinese, South Asian, black, Arab/West Asian, Filipino, South East Asian, Latin American, Japanese, Korean or other.

> The long version of the latest census brought dreaded Question 19 right on to my lap. This is the question that offers us 11 categories from which to choose a suitable label.

I had been dreading receiving this form because it would force me to consider the illegal act of not answering the question, thereby inviting a fine of $500.

I had been contemplating breaking the law because I had no wish to respond to a question that, based on published reports, confused race with ethnicity and nationality in attempting to quantify the number of visible minorities in Canada.

Reading the question reinforced my position, if only because the question generated so much confusion.

Some of the choices actually offered sub-categories that added language to the mix. For example, the category South Asian offered in brackets the descriptions East Indian, Pakistani, Punjabi and Sri Lankan.

Even if you ignored the historic anomaly of calling someone from India, like me, an *East* Indian, the question still equated the language Punjabi with someone who might have roots in the provinces called Punjab in India and Pakistan.

Similarly, black offered the examples of African, Haitian, Jamaican and Somali. How could our census drafters ignore the white "visible" minorities of Africa?

The categories for Arab/West Asian made the whole matter even more confusing. These were listed as Armenian, Egyptian, Iranian, Lebanese and Moroccan, ranging, as you can see, over a wide geographic region and including at least two continents.

If the presence of sub-categories added to the confusion, so did their absence.

Who, I wondered, would be covered by the Latin American label? If you were to use the Spanish language as the criterion, then you'd have to include North Americans like Mexicans in the category. But then where would that leave the Portuguese-speaking Brazilians?

Again, if you restricted Latin Americans to the South American continent, then where would you place Guyanese, large numbers of whom are of African or Indian origin?

I felt mildly grateful that Chinese, Japanese and Korean weren't classed under the label Asian, the word Canadians and Americans generally (mis)use to describe them.

But I did wonder where I'd put myself if I was Chinese, but not from China. If I was from Vietnam, I did have the choice of ticking off South East Asian, which offered Vietnamese as a sub-category, along with Cambodian, Indonesian and Laotian.

This raised the issue of motivation.

Would it really matter to the census people which label I chose for being Chinese? After all, all they wanted to know was whether I was a visible minority or not.

What difference would it make whether I ticked Chinese or put myself under South East Asian because I was an ethnic Chinese from Vietnam? I'd still fit the bill for the stereotype of "Asian," wouldn't I?

But if this was the intent of the question, then why offer separate categories for Chinese, South East Asian, Japanese and Korean?

Could it be, perhaps, that the census gurus felt that some categories are more homogeneous than others?

If so, it would explain why they chose not to qualify the label "white" even though they felt compelled to do so with the label "black."

At this stage, I found myself overwhelmed by all the contradictions and confusion of Question 19.

No matter how I looked at the question, skin color seemed to seep into the picture, raising the ugly spectre of racism.

Yet I felt uneasy about labelling the question racist because census officials have publicly stated the data it will yield will help advance the cause of employment equity.

Torn, I took what I hope was the honorable way out.

Beside the category called "other" I scribbled the word Canadian.

It's a label I'd be proud to defend in court.

On the margins

"**I** don't want to live here any more!"

The anguished wail seemed to come from behind me. As I turned to look, my foot slipped and I found myself floundering in the placid waters of the holy Indian river I had been standing in.

> It matters little whose side you're on or which party you support. Once an ethnic, always an ethnic. Once you arrive in the land of your dreams, all you find are nightmares.

Back on land, wet and muddy, I found myself cycling along a mountain road. A fellow cyclist wearing a Maple Leaf T-shirt pointed to the brilliant Caribbean sunset, lost his balance and disappeared into a snowbank.

From somewhere a train whistled and a loud chant broke out. Re-FU-gee! Re-FU-gee! Re-FU-gee!

It was too much for my busy brain. I woke up shaken but amused.

Or was I? I'm not sure.

It did seem rather silly, this terror-scored dream of conflicting images. Yet it contained enough disturbing echoes of real life to make me uneasy.

Once again I found myself tangled in the questions that

seem to dog every newcomer. When do you cease being an immigrant? What does it mean to be Canadian? When are you accepted as one? And what is the price of acceptance?

If the past year and our politicians are any guide, the answer seems loud and clear – *never!*

It matters little whose side you're on or which party you support. Once an ethnic, always an ethnic. Once in the land of your dreams, all you find are nightmares.

If you're not colluding with big business and screwing up the referendum results, then you're having too many babies or collecting too much welfare. Somewhere in between you're also a menace because you're not producing white babies.

The dust barely settles on one comment before there's a new one from someone else. Tories, Reformers, Péquistes and Bloquistes have all been guilty of stoking ethnic prejudice, all in the name of freedom and democracy. They've all been attacked and denounced, too. Often shrilly.

Last week it was the Liberals' turn. We had the sorry spectacle of Doug Young, the federal human resources minister, suggesting that a member of Parliament should find another country if he doesn't like Canada's immigration policy.

Young directed his anger at Oswaldo Nunez, the Chile-born Bloquiste. Young felt Nunez had no business preaching separatism because he was welcomed to Canada when he fled Chile as a refugee. Predictably, Young was denounced, although not by his boss, Prime Minister Jean Chrétien.

Predictably, the denunciations, like Chrétien's support, were made in the name of freedom and democracy. There was also some pious talk about liberty and human rights from all sides.

Almost all the denunciations, including a demand for an apology by Young in a *Gazette* editorial, came from Quebec, making the affair a national-unity issue.

Ironically, it was Quebec Premier Lucien Bouchard who injected the larger context. The comment by Young, Bouchard said, "means that Tremblay and Bouchard from Lac St. Jean and Saguenay would have the right to be elected in Ottawa and promote the sovereignty of Quebec but immigrants or recent citizens would be denied the right.

"It's a strange concept of citizenship."

Indeed. I'd stand up and cheer more loudly if the comment wasn't tainted by Bouchard and other sovereignists' record on tolerance of immigrants or their views.

But Bouchard is certainly right on the implications of Young's outburst.

By making place of birth and length of citizenship an issue, Young is making a mockery of the democratic standards of Canada that he claims to be defending.

The message he's sending out seems to be that the only good newcomer is one who agrees with the government. As any ethnic will tell you, that is similar to the message being sent out by the government in Quebec.

Both messages are dangerous because they don't distinguish between the policies of the government of the day and the ideals of the larger, diverse society they're members of but never wholly representative of. It may come as a shock to Young, but newcomers do share a passion for these ideals, whether they've been citizens for a year or for many generations. Often, it's this passion that brought them or their ancestors here in the first place.

All of which does little or nothing to resolve a newcomer's dilemmas. The path to full acceptance remains rocky as ever, whether you're stumbling into a river or cycling to some distant peak.

The most you can do is to keep on trudging.

Time for tea

Chai garam! Chai! Chai garam! Chai!

The tea vendors' hoarse calls would rise above the whoosh and clang of the train's brakes.

It wouldn't matter if it was 3 in the morning.

While there were always a few who could snore through the noisy stops at one of the hundreds of dimly lit railway stations that dot the Indian countryside, many others would stir themselves awake for a hit of the sweet, steaming, milky tea.

> The thick chai of my childhood is one that I can easily brew anywhere but never replicate entirely. Like any comfort food or ritual, the only standard is the one frozen in your consciousness at the moment it was born.

The quality and serving of the chai would reflect the size and location of the town.

At a major junction, the vendors would be pushing gleaming steel carts holding vats of chai and stacks of cups and saucers.

But for earthy taste and effect there was nothing to beat

the rural whistle-stops where often a single vendor would service an entire train, racing from window to window with a steaming kettle and disposable clay cups.

An enduring image that crops up as frequently as do trains in my sleep is that of a barefoot vendor in khaki shorts running beside a departing train, collecting coins for his chai until he is forced to pull up short as the platform ends.

That the vendor also ended up short on payments was a given. Time and cheapskates were deadly enemies.

But the next express would be due soon. Time to brew more chai.

As often as that vendor or his call runs through my mind, so do cravings for his chai.

These cravings surfaced last week on learning that a Health Canada study had found evidence that tea might help inhibit certain forms of cell mutations caused by cancer.

What joy, I thought even as I contemplated my chai cravings. Like all cravings, this one is personal and irrational. It doesn't mean that I don't like chai the way they brew and drink it in Catai, as China was called by the Venetian travellers.

Indeed, like millions of tea lovers, I can only be grateful to Emperor Sheng Nung, who is credited with discovering the gastronomic and medicinal potential of the tea leaves that legend says, accidentally blew into the water he was boiling under a tree in 2737 BC.

But for that chance discovery, the Japanese would have been unable to develop their famed Cha-no-yu ceremony once they began growing tea around 800 AD.

I can't think of a more seductive and soothing ceremony to draw cultural sustenance from. Art, aesthetics, philosophy and spiritual exploration are the integral and enduring components of an elaborate ritual that ends up enriching your life immeasurably.

But I was born in India, which only started growing tea in British colonial times. And for reasons I am ignorant of, the Indians began adding milk, sugar and aromatic spices to their tea.

Sure, they also contributed to, or took part in the afternoon tea tradition the British perfected in the 19th century.

But access to this world of silver and porcelain, scones and cucumber sandwiches, was clearly limited to the elite.

No, tea's appeal to the common Indian lay not in dainty rituals, but in its ability to stem appetite, while delivering the stimulus of caffeine and tannin.

Thus the basic brew they evolved is one of black tea leaves boiled with water, milk and sugar into a latte-like beverage.

For those who can afford it and certainly for all special occasions, cardamom, cloves, cinnamon and ginger are boiled along with the basic brew.

With or without spices, this tea can be found all over India, be it a deluxe palace hotel or a roadside shack.

This, then, is the thick chai of my childhood that I can easily brew anywhere but never replicate entirely. Like any comfort food or ritual, the only standard is the one frozen in your consciousness at the moment it was born.

If it were possible to re-create it with precision, it would instantly cease to be unique. It would also dampen the imagination even as it soured the spirit.

No, any comfort crutch has to be unbreakableto help you as much with your dreams as with reality.

How else can you gaze back at a forlorn vendor watching you recede from his life as a departing train gathers speed?

How else can you savor that milky, dark amber brew from a mug of clay? Inhale that cardamom steam reflected in the window of the train as it is swallowed up by the darkness?

Only by making each sip last a lifetime.

Bowing to the right

On a cool February morning, a young man is leaping blindly across ornamental hedges and flower beds, clutching a stolen pink rose in one hand and his spectacles in the other.

Pursuing him is a dishevelled caretaker, yelling and brandishing a thick bamboo stick.

Maybe because he's tired from his night shift or still groggy from having dozed off, the caretaker is unable to nab the rose snatcher.

> It wasn't that long ago that young men and women sang and marched for liberal values. Their counterparts today are showing alarming right-wing tendencies. Youth and idealism don't seem to go hand in hand any more.

"He almost got me," I remember my college chum Raman Suri saying later, stroking his goatee with some agitation.

"All I wanted was one rose, just one! For Cathy! For peace and love!"

Sounds corny today, but it seemed normal in 1968 in the central Indian city of Nagpur, where Suri and I were university students. Cathy had briefly entered our lives as one of seven students participating in something called Project India organized by the University of California at Los Angeles.

Cathy was the one with golden hair and blue eyes in a group that included a black woman and a Jewish man to reflect America's diversity.

But this typically Californian initiative had been lost on the mostly male audiences the group had attracted on Indian campuses.

All eyes always seemed to be on Cathy, not just because she personified the stereotype of a blonde California goddess, but also because she played guitar and sang.

The day before a smitten Suri became a petty thief, he had been spellbound by Cathy's rendering of *Blowin' in the Wind*.

So had I.

We were barely 18 and this was the first time we had heard this and the other folksongs of freedom and rebellion that the Americans had brought to our campus.

Stirred and shaken, we had talked up a storm that day – with Cathy and her gang and, later, just the two of us.

We didn't know it then, but we had become part of an international phenomenon.

All over the democratic world that year there was turmoil on college campuses. Sickened by the oppressive excesses of the previous generation, the young were rebelling.

But whatever their methods – not everyone followed the peaceful sit-in path – they all shared several important ideals.

Freedom and equality, truth and justice, fairness and compassion.

These were also the mantras that Suri and I took to chanting, long after Cathy had departed with her stolen rose.

Graduation, higher studies, job hunts and migration

would eventually disperse us, but in those heady days all we and other friends talked or marched about were these common ideals.

Looking back today and talking to old friends and new ones from that generation, I am struck by the durability and universality of the liberal values we sang and agitated for.

But comforting as this realization might be, it would be foolish to sink into a misty-eyed, baby-boomer stupor as a result of it.

This is because, everywhere you turn today, these cherished liberal values are under attack.

And – how's this for irony? – some of the most vocal assaults are coming from people under 25!

The first round was fired last week by the youth wing of the Quebec Liberal party.

Force able-bodied welfare recipients to do community service, the young Liberals proclaimed in a resolution adopted at a convention whose slogan was "The Revolution of a Generation."

Other "revolutionary" proposals endorsed included changes in the Quebec pension plan to ensure the young get their share at retirement.

The next salvo came from the youth wing of the federal Tories, a party that has ostensibly been struggling towards the middle since its disastrous performance in the last election.

The young Tories want a return to capital punishment, a 20-per-cent cut in personal taxes, an end to affirmative action, immediate deportation of refugees and immigrants who break the law, more restrictive social programs and boot camp for wayward teenagers.

A party official defended these proposals as symbolic of a natural tendency in youth to push for change. In a different time, such a naive defence would be laughed at.

But in today's political climate, it underscores the sinister reality of the declining importance of humanist values.

Until recently, these erosive attacks were coming from older generations and ultra right-wing groups.

That the disease is spreading to moderate groups is alarming enough. What is chilling is that it's blurring political lines and claiming younger victims.

It all makes you wonder what's blowing in the wind these days.

Certainly not the answers millions found barely three decades ago.